TECHNOLOGY DIFFERENCES
OVER SPACE AND TIME

TECHNOLOGY DIFFERENCES OVER SPACE AND TIME

Francesco Caselli

*This work is published in association with
the Centre de Recerca en Economia Internacional (CREI)*

PRINCETON UNIVERSITY PRESS
PRINCETON AND OXFORD

Copyright © 2017 by Princeton University Press
Published by Princeton University Press, 41 William Street,
Princeton, New Jersey 08540
In the United Kingdom: Princeton University Press,
6 Oxford Street, Woodstock, Oxfordshire OX20 1TR

press.princeton.edu

Jacket images courtesy of Shutterstock

Library of Congress Control Number 2016954450

ISBN 978-0-691-14602-7

British Library Cataloging-in-Publication Data is available

This book has been composed in Linux Libertine and Myriad for display

Printed on acid-free paper. ∞

Printed in the United States of America

1 3 5 7 9 10 8 6 4 2

A mio padre e a mia madre

CONTENTS

PREFACE

This book is based on part of my "CREI Lectures," given in Barcelona on June 16–18, 2010, at the end of a wonderful year-long visit to the Centre de Recerca en Economia Internacional (CREI). It was a great honor to be invited to give these lectures, and a splendid intellectual and personal experience to visit the Centre. For both, I am very grateful to Hans-Joachim Voth, Jaume Ventura, and Jordi Galí, and all the rest of the academic and nonacademic staff.

The material in the book extends ideas and updates results that I originally developed in published works written with John Coleman. To John goes my gratitude for a most fruitful (and ongoing) collaboration and friendship. My work with Jim Feyrer and Dan Wilson was also important in shaping ideas presented here, and I thank them too.

The book owes much to the dedication and resourcefulness of Jacopo Ponticelli, who did an enormous amount of work to help me create the original lectures, and Federico Rossi, who put an equally enormous effort into helping me turn the lectures into the book. Jacopo and Federico are also coauthors of the appendix on Mincerian returns.

It is customary for authors to thank their spouses for their forbearance during the gestation of a book. This book is too slender to have caused Silvana undue distress from any type of vacancy on my part. I still thank her, though, because there is not an idea in this book, or for that matter in anything I have done in the last 15 years, which she has not enriched with her intelligence and generosity.

TECHNOLOGY DIFFERENCES
OVER SPACE AND TIME

1

INTRODUCTION AND PRELIMINARIES

1.1 A Nontechnical Overview

Economies rely on a rich set of different inputs to produce goods and services. The composition of the pool of productive resources varies dramatically across countries. Some countries have a lot of workers with little education and few highly educated workers, while others have an abundance of workers with many years of schooling. Some countries have a copious supply of natural resources, relative to their stock of productive machinery and structures, while others are resource-poor but have plenty of equipment. Some countries have a lot of people and little capital; others have high capital-labor ratios.

Do these big differences in the composition of the bundle of factors of production lead to correspondingly large differences in the way productive activities are organized? Do productive units in a country tailor their mode of production, choice of technology, and management practices to the particular patterns of scarcity and abundance of inputs in their country? And if so, how? Do they choose production methods which make the most of the abundant factors—giving up, so to speak, on those that are relatively limited in their countries? Or do they rather focus on maximizing the contribution of the inputs in limited supply, to make up for their scarcity?

Similar observations, leading to similar questions, can be made for a given country over time. In the space of years and decades, the relative supply of skills, equipment and structure, natural resources, and even workers belonging to different demographic groups can change substantially. Do modes of production change and adapt in response to these changes? If so, does the adaptation compound the changes, by increasing the reliance of production units on the inputs that are becoming more abundant; or does it lean against the wind, by strengthening the ability to contribute to output of the factors of production that are becoming relatively scarce?

This book reports on my attempts to provide some answers to these questions. It shows that the mode of production—or the "production technology" in the words I use throughout the book—does vary systematically across countries, depending on their endowments of different factors of production. The production technology also changes over time, as factor supplies change. As to whether this adaptation favors the abundant or the scarce inputs, the answer turns out to be "it depends." When the factors becoming scarce are not that different, in terms of the role they play in the production process, from those becoming abundant, the production technology adapts to maximize the impact of the abundant factors. On the other hand, when the scarce factors are special and are difficult to replace using the abundant factors, the production technology mutates to bolster the contribution of the scarce inputs.

Admittedly, none of the above sounds particularly surprising. What is perhaps more surprising is that these patterns of technology adaptation are completely ignored by the vast majority of economists' thinking on how the organization of production varies across countries and (to a slightly lesser extent) over time. The dominant view, instead, is that some countries just have "better" technology than others, regardless of differences in factor endowment. Therefore, regardless of factor endowments, countries with "poor" technology should strive to copy as best they can the modes of organization they observe in countries with "good" technology. Similarly, over time technology mostly just gets "better," lifting the productivity of all factors equally. This pervasive, and rather boring, view of technology is incompatible with the evidence presented in this book: technology differences over space and time are much more interesting than most economists make them out to be!

1.2 A Slightly More Technical Overview

Economists characterize the relationship between a country's productive resources and its GDP by means of the *aggregate production function*. The aggregate production function can be used to answer two types of questions: (i) If country A has $x\%$ more of a given input (say, labor) than country B, by how much will country A's GDP exceed country B's (everything else being constant)? (ii) If country A experiences an $x\%$ increase in a given input between years t and $t + 1$, by how much will its GDP increase between the two years?

In empirical applications, economists have long noticed that production functions are not stable. Namely, the mapping from inputs onto outputs changes both across countries and over time. It is customary to refer to this instability as *technology differences* (across countries) and *technical change* (over time).

A long tradition of studies attempt to quantify the *pace* of technical change. This endeavor is usually refereed to as *growth accounting*. Growth-accounting exercises usually find technical change to be very important in driving changes in GDP over time. There is also a more recent, but now well-established, strand on quantifying the *magnitude* of technology differences across countries. These *development accounting* studies tend to find that technology differences are very important in determining differences in GDP. Both these sets of findings have profoundly influenced the way economists think about economic performance in the long run.

While these traditions have been effective at quantifying the extent of technology differences and technical change, they have arguably been less successful at characterizing their nature. In the vast majority of cross-country empirical applications technology is assumed to be *factor neutral.* Roughly speaking (and I will of course be more precise below) factor neutrality implies that if the efficiency with which a country uses one input is $x\%$ higher than the same input's efficiency in another, then the efficiency of all other inputs is also $x\%$ higher. Discussions of growth-accounting exercises are sometimes more nuanced and more aware of the possibility of technical change that is not factor neutral. Still, the methodology delivers a single quantitative measure of the pace of technical change, and is unsuited to characterize its nature. For this reason, growth-accounting results are still almost universally interpreted as if technical change was factor neutral.

This factor-neutral representation of technology is incompatible with various facts about factor prices across countries and time. As I will explain, if that view was correct, skill premia in (skill-) poor countries would be much higher, compared to premia in (skill-) rich countries, than they are; and skill premia couldn't have grown over time the way they have. Also, the capital share in income would vary much more across countries and over time than it does in the data. Changes in the experience premium over time are also inconsistent with neutral technical change.

This book uses these and other observations to show that technology differences and technical change are *factor biased*: they change not only the overall efficiency with which a country exploits its bundle of productive inputs, but also the relative efficiency with which different factors contribute to production. In fact, in some cases evidence shows that as the efficiency with which one input increases, the efficiency of another *decreases.* Allowing factor-biased technical change helps explain, among other things, why skill premia are very similar in poor and rich countries; why they have been growing over time; why the experience premium has not declined in response to the maturing of the baby-boom generation; and why the capital share in income is fairly constant both across countries and over time.

More specifically, I show that in richer countries the efficiency with which skilled labor is used relative to unskilled labor is greater than in poorer countries; similarly, the efficiency with which reproducible capital (equipment and structure) is used relative to natural capital (mineral deposits, land, timber, etc.) is higher in rich countries; also, when comparing the efficiency of an overall bundle of labor inputs (appropriately combining skilled and unskilled labor) and an overall capital input (which combines reproducible and natural capital), rich countries use labor relatively more efficiently. Furthermore, the absolute efficiency with which physical capital is used appears to be not lower, and may even be higher, in poor countries.

Over time, I document (like others before me) an increase in the relative efficiency of skilled labor. I also find an increase in the relative efficiency of older workers relative to younger ones (holding skills constant). Finally, in an echo of the corresponding result in the cross section, the efficiency with which physical capital is used has been declining over time.

I interpret these findings by means of a simple theoretical model of endogenous technological choice. In the model, firms choose from a menu of technologies (production functions). The key consideration turns out to be the degree of substitutability among factors. When two factors of production are highly substitutable, firms choose technologies that maximize the efficiency of the cheaper factor (at the expense of the efficiency of the more expensive factor). Instead, when two factors are poor substitutes, firms choose to maximize the efficiency of the expensive factor.[1]

To see how this framework sheds light on the empirical findings, consider skilled and unskilled labor. Rich countries have larger relative supplies of skilled labor, and hence skilled labor is relatively cheap there. Since skilled and unskilled labor are pretty good substitutes, firms in rich countries seek to make

---

[1] In this book I use the phrase "technology choice" to indicate a choice of parameters for the production function, i.e. a choice about the mapping between inputs and outputs. This is different from an alternative usage, encounterted particularly but not exclusively in the classic trade literature, where technology choice refers to a choice of a particular input (e.g. capital-labor) ratio.

the most of skilled labor, and end up picking technologies that imply a high relative efficiency of skilled labor compared to poor countries. Rich countries also have larger relative supplies of physical capital (broadly construed to include natural and reproducible capital) compared to labor (broadly construed to account for the larger proportion of skilled workers). But labor and capital are (thought to be) poor substitutes, so in this case rich countries choose technologies that bolster the relative efficiency of capital. The other empirical patterns uncovered can be interpreted along similar lines.

The factor-neutrality approach implies a Manichean view where some countries "get it right" and others "get it wrong." Countries either make the most of their skilled and unskilled labor, reproducible and natural capital, or they fail to use any of these efficiently. One implication is that poor countries should strive to reproduce rich countries' technological choices, regardless of their factor endowments and other determinants of optimal technology choice. The nonneutrality findings in this book, and in the research on which this book is based, point to a more nuanced picture. To be sure, firms in poor countries lag far behind the technology frontier to which rich-country firms have access. But technology transfer and adoption should be selective and tailored to local conditions.

1.3 Aggregate Production Functions

The central—indeed the only—analytical tool used in this book is the aggregate production function. The aggregate production function is a mapping from a country's input *quantities* to a country's *output*, and we express it as

$$Y_{ct} = F_{ct}(X_{1ct}, X_{2ct}, \ldots), \tag{1.1}$$

where Y_{ct} is aggregate output in country c in year t, X_{jct} is the quantity of input j used in production, and F_{ct} is the mapping in question. Note that the mapping carries subscripts c and t, indicating that the aggregate production function is *country and time specific*.[2]

The empirical counterpart of output Y_{ct} is gross domestic product (GDP). More specifically, when we are concerned with cross-country comparisons, we focus on GDP at purchasing power parity (PPP GDP). PPP GDP adds up the quantities produced of all final goods and services using a common set of prices (PPPs) as weights. When making comparisons over time, constant-price series must be used.

This book is about *how* F_{ct} varies across countries and over time. The *factor-neutral* case can be represented as

$$F_{ct}(X_{1ct}, X_{2ct}, \dots) = A_{ct}\tilde{G}(X_{1ct}, X_{2ct}, \dots). \qquad (1.2)$$

In this special case, the production functions F differ by, and only by, the multiplicative term A. There already is a lot of literature documenting that A contributes substantially to changes in GDP over time and very substantially to cross-country differences in GDP.[3]

Factor neutrality is a natural first step in investigating cross-country technology differences as well as technical change, but a glance at equation (1.2) clearly shows that it is highly restrictive. The book focuses on the following conceptual generalization:

$$F_{ct}(X_{1ct}, X_{2ct}, \dots) = G(A_{1ct}X_{1ct}, A_{2ct}X_{2ct}, \dots). \qquad (1.3)$$

In (1.3) the technology parameter A_{jct} augments factor j. A country (year) may have a relatively high value of one of the A_{jct}s

---

[2] I will not list here the conditions under which an economy admits an aggregate production function, but I do remind the reader that this list is exceptionally stringent. I am not aware of systematic attempts to assess the extent to which one can still count on aggregate production functions "approximately" holding when such assumptions are violated, as they certainly are. But everything in this book is subject to the (very big) caveat that something like an approximate production function still holds despite the certainty of violation of the theoretical underpinnings.

[3] A brief overview of growth accounting (which studies changes in A_{ct} over time) with references to the classic contributions can be found in Caselli (2008a). A brief overview of development accounting (across countries) is provided in Caselli (2008b) and a detailed one in Caselli (2005).

without having a proportionally high value of another. In other words, technology differences need not be factor neutral—though neutrality is admitted as a possible special case. The book is about asking whether—and if so how—the A_{jct}s vary across countries and over time.[4]

To this end, we must begin by identifying the list of relevant factors of production. I focus on four broad aggregates: unskilled labor, skilled labor, physical reproducible capital, and natural capital. The breakdown of the main factors of production into labor and capital is almost as old as economics, and the breakdown of labor into skilled and unskilled is also well established. The importance of accounting for reproducible and physical capital has recently been emphasized by Caselli and Feyrer (2007).

We must also specify a functional form for G. The book applies methods originally developed by Caselli and Coleman (2002, 2006) and Caselli (2005) which allow for identification of the A_{jct}s when the production function features constant elasticities of substitution (CES). Accordingly, in most of the book, I work with the following specification:

$$Y_{ct} = [(A_{Kct}K_{ct})^\sigma + (A_{Lct}L_{ct})^\sigma]^{1/\sigma}, \quad (1.4)$$

$$K_{ct} = [(A_{Nct}N_{ct})^\eta + (A_{Mct}M_{ct})^\eta]^{1/\eta}, \quad (1.5)$$

$$L_{ct} = [(A_{Uct}U_{ct})^\rho + (A_{Sct}S_{ct})^\rho]^{1/\rho}. \quad (1.6)$$

Hence, the production process is represented by a sequence of nested CES aggregators. Beginning from the bottom, unskilled labor U and skilled labor S are combined into an aggregate labor input L with elasticity of substitution $1/(1 - \rho)$. Similarly, natural capital N and reproducible capital M (M for machine) are combined into the aggregate K, with elasticity of substitution $1/(1 - \eta)$. Finally, labor and capital are aggregated with elasticity

---

[4] Clearly (1.3) remains restrictive in that it only admits technology differences of the factor-augmenting kind—there are no c or t subscripts to the function G.

$1/(1 - \sigma)$ to produce output. Technology differences are captured by differences in the factor-augmenting terms A_{Uct}, A_{Sct}, A_{Nct}, A_{Mct}, A_{Kct}, and A_{Lct}, which are the object of this study.[5]

The advantage of the nested-CES structure is twofold. First, it keeps the number of parameters (other than the augmentation factors A) to a minimum, i.e. the three elasticities of substitution. Second, as we will see, it allows for breaking up the problem of identifying the relative efficiency of any two factors into stages, i.e. first between skilled and unskilled labor, then between reproducible and natural capital, and only then between labor and capital. Admittedly other nestings are in principle possible, and little in the literature offers guidance on the most appropriate one. I have chosen the one in (1.4)–(1.6), as it is the most consistent with traditions emphasizing the distinction between skilled and unskilled labor, and labor and capital. Perhaps more important, the existence of these traditions provides (some) information on the plausible values of the corresponding elasticities of substitution.

With a slight modification (discussed below in section 1.8), the CES aggregates in (1.4)–(1.6) nest the Cobb-Douglas case as a special case. Macroeconomists often use the Cobb-Douglas assumption, particularly for (1.4), on the ground that the capital share is constant in the United States. The historical trendlessness of the capital share in the United States, however, can of course be replicated by CES models with the "right" time series behavior of the effective supplies of capital and labor (i.e. $A_K K$ and $A_L L$). Furthermore, there is clear evidence of substantial fluctuations in the capital shares of many countries other than the United States, and even there in recent years [e.g. Oberfield and Raval (2012), Elsby et al. (2013), Neiman and Karabarbounis (2014)].

---

[5] In chapter 6 I add a furhter level of nesting "under" equation (1.6), where U and S are further broken down by the amount of experience.

1.4 Factor Bias

It is useful to establish a terminology to characterize particular patterns of variation of technology across countries and over time. To do so, I build on the terminology that was developed to characterize technical change over time, and extend it to the cross-country context.

Consider again an aggregator of the form

$$X = \left[(A_1 X_1)^\zeta + (A_2 X_2)^\zeta \right]^{1/\zeta} . \tag{1.7}$$

In the time series, it is customary to say that technical change is *factor-i augmenting* if A_i increases over time. Furthermore, technical change is said to be *biased toward factor i* if $(A_i/A_j)^\zeta$ increases over time.[6]

To see the rationale for the definition of factor bias note that

$$\frac{MP_i}{MP_j} \propto \left(\frac{A_i}{A_j} \right)^\zeta \left(\frac{X_i}{X_j} \right)^{\zeta-1} ,$$

where MP_i (MP_j) is the marginal product of factor i (j). Hence, technical change is biased toward factor i if it increases the relative marginal productivity of factor i when relative factor quantities are held constant. In recent years the idea of factor bias in technical change has played a prominent role in attempts to explain changes in the wage structure [e.g. Katz and Murphy (1992), Acemoglu (1998, 2002), Autor, Katz, and Krueger (1998), Katz and Autor (1999), Caselli (1999), Goldin and Katz (2008)].

In a cross section of countries, similar definitions are possible if we replace time with a suitable criterion to order observations. The natural criterion is income per worker. Hence, we will say that technology differences across countries are factor-*i* augmenting

---

[6] The definitions of factor augmenting, neutral, and biased technical change go back to Hicks (1939).

if A_i is higher in countries with higher GDP. Furthermore, technology differences across countries are biased toward factor i if $(A_i/A_j)^\varsigma$ is higher in countries with higher GDP.[7]

1.5 Alternative Representation

It is immediate that an alternative representation for an aggregator of the form (1.7) is

$$X = \Omega_1 \left[(X_1)^\varsigma + \Omega (X_2)^\varsigma \right]^{1/\varsigma}, \tag{1.8}$$

where the mapping is

$$\begin{aligned} \Omega_1 &= A_1 \\ \Omega_2 &= \left(\tfrac{A_2}{A_1} \right)^\varsigma . \end{aligned} \tag{1.9}$$

In words, we can work with aggregators that are specified in terms of the augmentation coefficients of both inputs or in terms of one augmentation coefficient and one factor-bias coefficient. In the book, I will exploit this representational equivalence extensively.

1.6 Plan for the Book

The book is divided into three parts.

Part I is the "across-countries" part. In chapters 2 and 3 I will use the specification in (1.8) and (1.9) for equations (1.6) and, respectively, (1.5), to identify the *factor bias* (if any) in labor and capital aggregation. In other words, in these chapters I (drop time subscripts and) estimate A_{Sc}/A_{Uc} and, respectively, A_{Mc}/A_{Nc}, and characterize how they vary across countries—particularly as a function of GDP. While these chapters produce estimates of the ratios A_{Sc}/A_{Uc} and A_{Mc}/A_{Nc}, they do not pin down the absolute

---

[7] For a precedent on replacing the time index with a country's ranking in the world income distribution see Hall and Jones (1996).

levels of A_{Uc} and A_{Nc}. As mentioned, I find that both $(A_{Sc}/A_{Uc})^\rho$ and $(A_{Mc}/A_{Nc})^\eta$ are positively correlated with income per worker.

In chapter 4 I turn to equation (1.4), which I keep in its original form. Substituting from equations (1.6) and (1.5), in their alternative form, we have

$$Y_c = [(A_{Kc}A_{Nc}\tilde{K}_c)^\sigma + (A_{Lc}A_{Uc}\tilde{L}_c)^\sigma]^{1/\sigma}, \qquad (1.10)$$

where

$$\tilde{K}_c = \left[(N_c)^\eta + \left(\frac{A_{Mc}}{A_{Nc}} M_c \right)^\eta \right]^{1/\eta}, \qquad (1.11)$$

$$\tilde{L}_c = \left[(U_c)^\rho + \left(\frac{A_{Sc}}{A_{Uc}} S_c \right)^\rho \right]^{1/\rho}. \qquad (1.12)$$

These substitutions reveal that, in a system of nested CES functions, it is not possible to separately identify the augmentation coefficient of all inputs at all levels of the nesting. Accordingly, chapter 4 focuses on estimating the augmentation coefficients

$$\tilde{A}_{Kc} = A_{Kc}A_{Nc},$$
$$\tilde{A}_{Lc} = A_{Lc}A_{Uc}.$$

We can think of these coefficients as augmentation coefficients for "natural capital equivalents" $\tilde{K}$, i.e. the capital input expressed in efficiency units of natural capital, and "unskilled-labor equivalents" $\tilde{L}$, or the labor input in efficiency units of unskilled labor. My finding is that $\tilde{A}_{Lc}$ is increasing in income per worker, while $\tilde{A}_{Kc}$ is either unrelated or perhaps even slightly decreasing in income per worker.

Part III is the "over time" part. In chapter 6 I extend the definition of the aggregate labor input in (1.6) to further break down the skilled and unskilled labor aggregates by experience. This results in an additional layer of CES nesting. I then show that the efficiency of experienced skilled (unskilled) workers increases over time in the United States relative to the efficiency of inexperienced skilled (unskilled) workers. I also look at the evolution over time of the relative efficiency of skilled workers to unskilled workers and confirm the skilled-biased technical change (SBTC) result.

Chapter 7 extends the time series analysis to a panel of Organization for Economic Cooperation and Development (OECD) countries and investigates both skill bias in technical change and the evolution of the efficiency of labor relative to capital. The analysis confirms that SBTC is a global phenomenon. More originally, I find that in almost all OECD countries $\tilde{A}_{Kct}$ has been declining over time.

Between the empirical parts I and III, in part II I pause for a theoretical interlude. I present a model of endogenous technological choice and use it to interpret the results of part I. At the end of part III I also return to the theoretical model to interpret that part's results.

1.7 Relation to Previous Work

All of the empirical results presented in this book are previously unpublished, in the sense that, at a minimum, they are obtained with data that have been updated with the most recent available sources. In most cases, however, I also extend previous work in various conceptual and methodological directions.

The analysis of skilled bias across countries in chapter 2 is based on Caselli and Coleman (2006). The data used in that paper refer to the year 1985 and covers a cross section of 52 countries. Here I report updated results on two cross sections: 1995 (66 countries) and 2005 (34 countries). I also improve very substantially on the methodology to construct the skilled and unskilled labor aggregates and to estimate the skill premium, which is a key input in backing out relative efficiencies.

The analysis of the relative efficiency of reproducible and natural capital in chapter 3 is novel to this book, though it is inspired by my work with Jim Feyrer [Caselli and Feyrer (2007)], which shows the importance of accounting for natural capital in estimating aggregate returns to capital across countries.

Chapter 4, which investigates how the efficiency of capital and labor (both broadly construed) varies across countries, updates the corresponding analysis in section 7 of Caselli (2005). There I looked at 96 countries in 1996. Here I present estimates for 1995 (but with revised data) and 2005. I also measure both the capital and the labor aggregates differently. In particular, I include natural capital in the former, and allow for imperfect substitution between skilled and unskilled labor in the latter. I also present extensions not present in Caselli (2005) in which the health status of the population and cognitive skills are allowed to contribute to differences across countries in the labor endowment.

In chapter 6, the study of experience bias in the United States is novel to this book, though it is heavily indebted to the original investigation of this theme in Katz and Murphy (1992). So is the study of skill bias, which, however, is methodologically closer to Caselli and Coleman (2002). The exercise on the OECD panel in chapter 7 is novel to this book.

The two-factor theoretical model of endogenous technology choice in part II is from Caselli and Coleman (2006). The extension to four factors is novel to this book.

1.8 A Note on "Share Parameters"

Before starting, a quick note to reassure readers who find my representation of aggregators of the form (1.7) unfamiliar. It would indeed be more rigorous to write

$$X = [\omega(\tilde{A}_1 X_1)^\zeta + (1 - \omega)(\tilde{A}_2 X_2)^\zeta]^{1/\zeta}, \qquad (1.13)$$

where ω is customarily referred to as the "share parameter." This specification is more accurate because it allows retrieval of the Cobb-Douglas specification as the limiting case when $\zeta \to 0$. In this limit, ω and $1 - \omega$ are indeed the factor shares (hence the terminology).

The factor shares are omitted here exclusively for ease of notation. The reader should simply keep in mind that any estimate of A_1 (A_2) presented in the book is really an estimate of $\omega^{1/\zeta} \tilde{A}_1$ ($(1 - \omega)^{1/\zeta} \tilde{A}_2$), as is easily verified by comparison of (1.7) with (1.13).

PART I

TECHNOLOGY DIFFERENCES
ACROSS SPACE

2

SKILLED AND UNSKILLED LABOR

2.1 Estimating the Skill Bias

In this chapter I focus on equation (1.12) and use it to assess how A_{Sc}/A_{Uc} varies across countries. The methodology to infer A_{Sc}/A_{Uc} for country c is very simple. Define W_{Sc} as the wage rate for skilled labor and W_{Uc} as the wage rate for unskilled labor. Assume now that labor markets approximate conditions of perfect competition. Then the system (1.10)–(1.12) implies[1]

$$\frac{W_{Sc}}{W_{Uc}} = \left(\frac{A_{Sc}}{A_{Uc}}\right)^{\rho} \left(\frac{S_c}{U_c}\right)^{\rho-1}. \tag{2.1}$$

The interpretation of this equation is that the relative wage of a skilled worker is decreasing with the relative supply of skills. However, for a given supply of skills the relative wage also depends on the relative efficiency with which skills are used. If skilled and unskilled labor are relatively good substitutes ($\rho > 0$), an increase in the relative efficiency of skills increases the relative marginal productivity of skills and boosts the skill premium. On the other hand, an increase in A_s/A_u also increases the *effective* relative supply of skills. If skilled and unskilled labor are relatively poor substitutes ($\rho < 0$), this relative supply effect dominates, and the skill premium declines in response to an increase in A_s/A_u.[2]

---

[1] In fact equation (2.1) holds for any aggregate production function of the form $Y = F(\tilde{L}, \dots)$, where $\tilde{L}$ is given by (1.12).

[2] Of course this is a partial equilibrium discussion. In general equilibrium, A_s/A_u is endogenous to L_s/L_u, as I show in part II.

Equation (2.1) implies that the unobservable quantity $(A_{Sc}/A_{Uc})^\rho$ can be inferred from data on the following (potential) observables: (i) the relative supply of skills S_c/U_c; (ii) the skill premium W_{Sc}/W_{Uc}. In addition, (iii) one has to calibrate the elasticity-of-substitution parameter ρ. I take up these three tasks in the next three sections.

It is important for this methodology that relative wages are informative about relative marginal productivities. If developing countries had more egalitarian labor market institutions, the observed skill premium in these countries would underestimate the difference between the marginal productivity of skilled and unskilled labor, potentially leading to a spurious evidence of skill bias. Of course, however, it is well known that—if anything—social and political pressures for containing wage dispersion are much more severe in rich than in poor countries (with the possible exception of the United States), so if anything this type of measurement error will bias the results against a finding of skill bias.

The methodology allows A_u/A_s to vary across countries, while ρ is constant, much as in the skilled-biased technical change literature. A certain amount of arbitrariness is of course present in the choice of which parameters vary, and which don't, across countries. This arbitrariness is inescapable: changes in ρ cannot be separately identified from changes in A_s/A_u, as is shown in the classic paper by Diamond, McFadden, and Rodriguez (1978).[3]

2.2 Estimating the Relative Supply of Skills

The key source of raw data to build measures of skilled and unskilled labor supply is a data set collected by Barro and Lee (2013), covering 146 countries at five-year intervals, from 1950 to 2010. The data set is best known for its variable "average years of

---

[3] It would, of course, be possible to fix A_S/A_U, and let ρ vary across countries. See Duffy and Papageorgiou (2000) for an effort in this direction.

schooling," which is an estimate of the number of years of education received by the representative worker. This variable has played a prominent role in the development-accounting literature discussed in the introductory chapter. In this study, however, I focus on a different set of variables from the data set, namely the share of individuals with different levels of schooling in the working-age population (proxied as the population over 15 years of age).

In particular, for each country and year, Barro and Lee report the proportion with (1) no education; (2) some primary schooling; (3) primary schooling completed; (4) some secondary schooling; (5) secondary schooling completed; (6) some college; (7) at least a college degree.

The first task in turning the seven achievement categories of Barro and Lee into an unskilled and, respectively, skilled aggregate is to choose an education threshold for "skilled." As explained later, the most credible available estimates of ρ use "secondary schooling completed" as the lowest *skilled* group. Accordingly, I classify groups 1–4 as unskilled and groups 5–7 as skilled.[4]

The next task is to decide how to aggregate the achievement subgroups within the unskilled and, respectively, skilled set. Because of lack of information on the patterns of substitutability within the unskilled and the skilled set, respectively, we will assume that subgroups 1–4 are perfect substitutes for each other, and so are groups 5–7. Hence, the unskilled and skilled aggregates take the forms

$$U_c = \sum_{j=1}^{4} e^{\beta_j} l_{jc} \tag{2.2}$$

$$S_c = \sum_{j=5}^{7} e^{\beta_j} l_{jc}, \tag{2.3}$$

---

[4] Needless to say it would be interesting to allow for finer classifications, with more than two skill groups. In fact, ideally one would treat all seven skill groups as imperfect substitutes. However, the microeconomic information necessary to calibrate a more complex labor aggregator is not currently available.

where l_{jc} is the share of achievement group j in the working-age population, $j = 1, \ldots, 7$.

The coefficients β_j measure relative endowments of efficiency units for workers with more or less education, *within* the unskilled and skilled aggregate, respectively. In particular, without loss of generality we can set

$$\beta_1 = \beta_5 = 0,$$

so that, for $j = 1, \ldots, 4$, β_j measures the endowment of efficiency units *relative* to a worker with no schooling, and, for $j = 5, \ldots, 7$, it measures the endowment of efficiency units *relative* to a worker who completed high school. In other words, our unskilled and skilled subaggregates are measured in units of workers with no schooling and, respectively, with a high school degree. Importantly, from this normalization it follows that the empirical counterpart of the skilled and unskilled wages W_{Sc} and W_{Uc} are the wages paid to workers who have completed high school and the workers who have no schooling, respectively.

The final task in building the U_c and S_c aggregates is thus to calibrate the β_js. Plugging (2.2) and (2.3) into (1.12), and using the last observation of the previous paragraph, we find that

$$W_{jc} = W_{Uc} e^{\beta_j} \qquad j \leq 4 \tag{2.4}$$

$$W_{jc} = W_{Sc} e^{\beta_j} \qquad j > 4, \tag{2.5}$$

where W_{jc} is the wage rate for a worker belonging to subgroup j, and W_{Uc} and W_{Sc} are functions of the relative labor endowments U_c and S_c. This suggests that the β_js can be estimated from individual-level wage and education data.[5]

In particular, suppose that for a certain country c we had data on a representative sample of workers, indexed by i, and

---

[5] The construction of overall labor inputs by aggregating subgroups, using relative wages as weights, of course has a long tradition in growth accounting. Here I am adapting this approach to a world where there could be two imperfectly substitutable labor-input bundles, with the further possibility of nonneutral technological differences.

belonging to the various attainment groups j. Then we could identify the βs in the previous equations by the two regressions:

$$\log\left(W^i_{jc}\right) = \log W_{Uc} + \sum_{j=2}^{4} \beta_j D^i_{jc} + \varepsilon^i_{jc} \quad j \leq 4 \qquad (2.6)$$

$$\log\left(W^i_{jc}\right) = \log W_{Sc} + \sum_{j=6}^{7} \beta_j D^i_{jc} + \varepsilon^i_{jc} \quad j > 4. \qquad (2.7)$$

In these regressions, W^i_{jc} is the wage of worker i belonging to achievement group j in country c, D^i_{jc} is a dummy variable that takes the value 1 if worker i belongs to achievement group j, ε^i_{jc} is an error term, and $\log W_{Uc}$, $\log W_{Sc}$, and the βs are parameters to be estimated (with the βs being the parameters of interest).[6]

Equations (2.6) and (2.7) are standard Mincerian log-wage equations, except that rather than measuring education with a single cardinal variable (years of schooling), we measure it via achievement dummies. I run the Mincerian-like regression separately for workers belonging to the unskilled and the skilled subgroups, because the model implies that the intercepts should be different for these two samples. Of course it would also be possible to retrieve the βs from a regression pooling all workers, by applying appropriate adjustments to the coefficients.[7]

---

[6] Here and elsewhere I adopt the convention that superscripts index individual workers while subscripts (other than c) will continue to denote achievement groups (subscript c continues to denote countries).

[7] Suppose we run the regression

$$\log\left(W^i_{jc}\right) = a_c + \sum_{j=2}^{7} b_j D^i_{jc} + \varepsilon^i_{jc}.$$

Clearly we have $a_c = \log\left(W_{Uc}\right)$ and $b_j = \beta_j$ for $j \leq 4$. For $j > 4$ we have

$$\log\left(W_{Sc}\right) + \beta_j = a_c + b_j,$$

and so

$$\log\left(W_{Sc}\right) = a_c + b_5$$
$$\beta_j = b_j - b_5 \quad j = 6, 7.$$

An important feature of regressions (2.6) and (2.7) is that, by assumption, the β_js do not vary across countries. This is in keeping with our maintained assumptions that technologies differ across countries only by the augmentation factors to skilled labor, unskilled labor, and natural and reproducible capital.

Because the β_js do not vary across countries, and because the intercepts are not of present interest, it is enough for our purposes to estimate (2.6) and (2.7) on data from a single country. For convenience, I use the United States. In particular, I use the Current Population Survey (CPS), which is widely regarded as a satisfactorily representative sample of American workers, including information on earnings, schooling, and other covariates. One shortcoming is that, since 1992, the variable describing educational attainment in the CPS does not map adequately into the seven achievement subgroups of Barro and Lee. Hence, I use data from 1991, which is the last year in which such a mapping is easily performed. The regressions are run on a sample including only white males, and control for a full set of age dummies. The results are displayed in table 2.1. Primary education confers approximately a 40% productivity increase over no schooling, and reaching secondary schooling a further 20%. Completing college increases productivity by about 50% over completing secondary schooling.[8]

With these estimates of the βs, our estimation of the labor aggregates in (2.2) and (2.3) is complete. Table 2.2 reports some summary statistics from the cross-sectional distribution of S_c/U_c relative to the United States [i.e. the distribution of $(S_c/U_c)/(S_{US}/U_{US})$] for the 146 countries in Barro and Lee (2013) for

---

[8] As in the Mincerian literature, the causal claims in the text should be taken with proper skepticism arising from the usual concerns with omitted variable bias. Nevertheless, in the Mincerian literature OLS and IV estimates of returns to schooling have generally been relatively close to each other, presumably because the upward bias conferred by the omission of unobserved ability roughly cancels out with the downward bias from measurement error. The causal intepretation is therefore less unwarranted than appears at first sight.

TABLE 2.1. Efficiency Units by Attainment Group

Unskilled		Skilled	
No schooling	0	Completed secondary	0
Some primary	0.32	Some college	0.14
Completed primary	0.38	College and more	0.46
Some secondary	0.56		

Coefficients from equations 2.6 and 2.7. CPS data.

TABLE 2.2. Summary Statistics for S_c/U_c

Year	Obs	Min	P10	P50	P90	Max	Corr w/Y
1995	146	0.003	0.016	0.078	0.253	1	0.38
2005	146	0.004	0.013	0.084	0.335	1	0.43

Relative to USA. Px = xth percentile. Y is income per worker.

the years 1995 and 2005. The choice of years is dictated by the availability of data necessary to estimate skill premia, as explained below.

In both decades, the variation in the relative supply of skills is enormous, with all countries below the median having less than 10% of the relative supply of skills of the United States and, even at the 90th percentile, still only having between a quarter and a third of the relative supply of skills of the United States, the country with the largest relative supply of skills in both subperiods. Between 1995 and 2005 the relative supply of skills catches up somewhat in the top half of the distribution, but not in the bottom half.

The table also reports correlations with income per worker, from version 7.1 of the Penn World Tables [Heston et al. (2012)], again relative to the United States. Not surprisingly these correlations are positive and indeed quite high. The relationship between the relative supply of skills and income is further illustrated in figure 2.1. In the figure, both axes are in log scales, but the labels correspond to the absolute values. Each country is represented by its three-letter World Bank code.

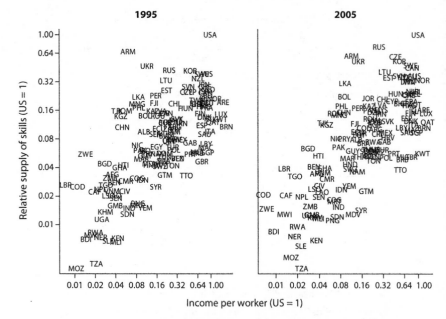

Figure 2.1. Cross-Sectional Distribution of the Relative Supply of Skills

2.3 Estimating the Skill Premium

Because S_c and U_c are measured in units of workers with no schooling and, respectively, workers with high school completed, the empirical counterpart of the skill premium W_{Sc}/W_{Uc} is the premium conferred by completing high school relative to never having attended school. Unfortunately, there is no readily accessible data set reporting the high school to no-schooling premium for a wide variety of countries. To construct such a data set, one would have to get hold of country-specific microeconomic data and re-estimate equations (2.6) and (2.7) (or the equivalent single-equation version) for each country in the sample. For each country, the (log of the) ratio W_{Sc}/W_{Uc} would be given by the difference in the two intercepts (or the coefficient on high school completed when the omitted category is no schooling).

Fortunately, a shortcut that provides an alternative to this immense task is available. As I describe below, it *is* possible to assemble a cross-country data set reporting *Mincerian returns*, or coefficients on years of schooling in regressions for the log wage. Conditional on our production model, and given knowledge of the distribution of workers by achievement group, it turns out to be possible to infer the skill premium from the Mincerian return, as I now show.

Consider a microeconomic data set, from a particular country, with information on years of schooling s^i and wages W^i for n workers, again indexed by i.[9] On this data set, we run the Mincerian regression

$$\log W^i = \alpha + bs^i + \varepsilon^i.$$

The coefficient b is the Mincerian coefficient. Using the OLS formula, b is

$$b = \frac{\sum_i \left(\log W^i - \mu_{\log W}\right)(s^i - \mu_s)}{\sum_i (s^i - \mu_s)^2},$$

where

$$\mu_s = \frac{1}{n}\sum_i s^i$$

and

$$\mu_{\log W} = \frac{1}{n}\sum_i \log (W^i).$$

Plugging in from equations (2.6) and (2.7), we can rewrite the Mincerian coefficient as

$$b = \frac{\sum_{j \leq 4}(\log W_U + \beta_j)(s_j - \mu_s)l_j + \sum_{j>4}(\log W_S + \beta_j)(s_j - \mu_s)l_j + \sum_i \varepsilon_i(s_i - \mu_s)}{\sum_j (s_j - \mu_s)^2 l_j},$$

(2.8)

---

[9] In the equations in this section I drop the country subscript as I work exclusively with within-country data.

where s_j is years of schooling of attainment group j. Clearly in rewriting the expression for the Mincerian coefficient this way I am relying heavily on treating years of schooling as a discrete variable, as implied by the structure of my data. Assuming that the error term ε_i is uncorrelated with years of schooling s_i, the last term in the numerator vanishes.[10] After some algebra (see Appendix A), the last expression can be shown to imply

$$\log W_S - \log W_U = \frac{b \sum_j (s_j - \mu_s)^2 l_j - \sum_j \beta_j (s_j - \mu_s) l_j}{\sum_{j>4} (s_j - \mu_s) l_j}.$$

(2.9)

This formula implies that it is possible to recover the skill premium from (i) the Mincerian return b (as already indicated); (ii) a measure of years of schooling for each of the seven attainment subgroups, $s_j, j = 1, \ldots, 7$; (iii) the shares of each subgroup in the labor force, $l_j, j = 1, \ldots, 7$; and (iv) the relative productivity parameters $\beta_j, j = 1, \ldots, 7$.

We obviously have item (iii) for a large cross section of countries, as discussed in the previous section. In that section we also constructed the parameters of item (iv). In the remainder of this section I discuss sources for (i) and (ii).

With Jacopo Ponticelli and Federico Rossi I created a new cross-country data set of Mincerian returns in the spirit of Psacharopoulos (e.g. 1994) and Bils and Klenow (2000). In particular, we undertook a broad search of the academic and policy literature on schooling and labor market outcomes, to extract estimates for b for as many countries as possible. This search yielded 81 observations for the period 1989–1999, for 78 of which we have the complementary data from Barro and Lee; and 75 observations (not necessarily the same countries) for the period subsequent to 2000, for 69 of which we also have Barro-Lee data.[11] All of these

---

[10] As already discussed in footnote 8, the assumption that ε_i and s_i are uncorrelated is very strong, but some solace can be found in the similarity of OLS and IV estimates.

[11] We collected up to one estimate per country per subperiod.

estimates, together with their sources, are reported in Appendix B, which also provides methodological details.[12]

The other item still required for estimating the skill premium using formula (2.9) is an estimate of the duration of each attainment level in each country. Data on duration of primary and secondary schooling are from the World Development Indicators (WDI), while data on duration of higher education are from Cohen and Soto (2007). Countries not covered by these sources are assigned the average durations of their "macro region" (as defined by the World Bank). For each level of education the fraction of the population that does not complete each level is assigned half the years of schooling of the full duration of that level.[13]

The estimated skill premia from the procedure described above feature three very large outliers, which I will omit from the subsequent analysis.[14] On the other hand, I can add a few direct estimates of W_{Sc}/W_{Uc} from log-wage equations specified in terms of achievement dummies. These were found in the course of the Mincerian literature search. With these subtractions and additions, I have 82 observations for the skill premium in 1995 and 84 in 2005.

Summary statistics for skill premia W_S/W_U (relative to the United States) are reported in table 2.3. Once again, cross-country variation in skill premia is tremendous, and there is some indication that the dispersion is growing over time. Also as expected, skill premia are lower in rich countries, where the relative supply

---

[12] A subset of these estimates is also reported in Caselli and Ciccone (2013).

[13] The duration data refer to 1995 in WDI (2012) and to various years (depending on the country) in Cohen and Soto (2007). It would be desirable to use duration data from the years when the average worker attended school. While in principle this could be done with the WDI data (since they include duration from 1970 on), Cohen and Soto report a single observation for each country. In any case the variation over time within countries is extremely small, so this is unlikely to bias the result. A more serious concern is that we are treating a given level of attainment, e.g. secondary completed, as conferring skills that are independent of the number of years required to reach that level, which varies (somewhat) across countries.

[14] The outliers are Jamaica in the 1990s and Rwanda in both decades.

TABLE 2.3. Summary Statistics for W_{Sc}/W_{Uc}

Year	Obs	Min	P10	P50	P90	Max	Corr w/Y
1995	82	0.34	0.67	0.95	1.83	4.01	−0.21
2005	84	0.17	0.47	0.82	1.90	5.51	−0.15

Relative to the United States. Px = xth percentile. Y is income per worker.

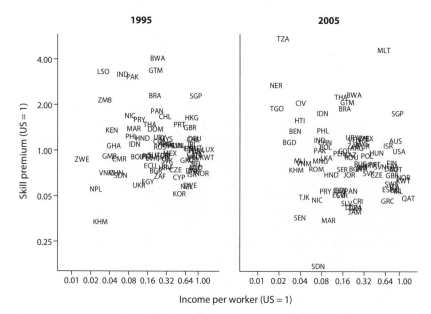

Figure 2.2. Cross-Country Distribution of the Skill Premium

of skills is larger (as we saw in the previous section). The actual skill premia are plotted against income per worker in figure 2.2.[15]

---

[15] The procedures followed here to build the relative supply of skills S/U and the skill premium W_S/W_U differ from those in Caselli and Coleman (2006). Caselli and Coleman proxy the β_js by $b \cdot \Delta s_j$, where Δs_j is the extra years of schooling of attainment group j relative to the benchmark group for group j (no schooling or high school completed); and they proxy W_S/W_U by $b \cdot (s_5 - s_1)$. Clearly the procedure in the text is more faithful to the underlying theoretical model. Having said that, the results do not seem very sensitive to these methodological differences.

2.4 Calibrating the Elasticity of Substitution

The last input required in order to back out the relative augmentation coefficients $(A_{Sc}/A_{Uc})^\rho$ from equation (2.1) is a calibrated value for the elasticity of substitution parameter ρ. Several authors have estimated the elasticity of substitution between skilled and unskilled labor, as reviewed by Autor et al. (1998). Few if any estimates lie outside the interval [1,2], and a majority cluster around 1.4 or 1.5. The most credible estimate is probably the one due to Ciccone and Peri (2005), who use variation across cities in the relative skill supply instrumented with compulsory-schooling laws. They set high school completed as the threshold for skill (hence my choice to do the same), and obtain an estimate of 1.5, corresponding to a value for ρ of 1/3.

In part II I present a model of endogenous technology choice where the skill-augmentation coefficient $(A_{Sc}/A_{Uc})^\rho$ depends on the relative factor supply, S/U. If the As are endogenous to factor supplies, changes in relative factor prices will reflect *both* the exogenous parameter that maps into the elasticity of substitution *sensu stricto* (ρ in the case of the present discussion) *and* the endogenous changes in factor-augmenting coefficients. Since elasticities of substitution are (typically) estimated on the response of relative factor prices to relative factor quantities, an important caveat to my calibration strategy is that estimates of the elasticity of substitution in the literature may not be clean estimates of ρ. The severity of this problem depends on the speed of adjustment of the As to the change in factor supplies, relative to the horizon over which the factor-supply changes used to estimate ρ take place. My conjecture is that changes in technology are relatively sluggish, so published estimates should be closer to the "short-run" elasticity $1/(1-\rho)$ than they are to the "full" elasticity, which also captures technology change.

2.5 The Skill Bias in Technology across Countries

Table 2.4 and figure 2.3 present the key empirical results of this chapter. Table 2.4 presents summary statistics from the cross-country distribution of $(A_{Sc}/A_{Uc})^\rho$. The cross-country heterogeneity in $(A_{Sc}/A_{Uc})^\rho$ is enormous: the 90th percentile exceeds the 10th percentile by a factor of 4 in the 1990s and by a factor of 6 in the 2000s. This implies an emphatic rejection of the view that technology differences are factor neutral. The relative efficiency with which skilled and unskilled workers are used varies massively across countries.

The table also shows the correlation between the relative augmentation coefficients $(A_{Sc}/A_{Uc})^\rho$ and income per worker, and their relationship is plotted in figure 2.3. Clearly, there is a strong skill bias in technology differences across countries: skill-abundant countries use skilled labor relatively more efficiently.

It is useful and instructive to investigate which features of the data give rise to this result. Begin by rewriting (2.1) in logs:

$$\log\left(\frac{W_{Sc}}{W_{Uc}}\right) = \rho \log\left(\frac{A_{Sc}}{A_{Uc}}\right) + (\rho - 1)\log\left(\frac{L_{Sc}}{L_{Uc}}\right). \quad (2.10)$$

Since $\rho = 1/3$, if relative augmentation coefficients were uncorrelated with relative skill supplies a regression of log wage premia on log relative skill supplies should yield a coefficient of around $-2/3$. Now consider figure 2.4. In this figure I plot $\log(W_{Sc}/W_{Uc})$ against $\log(L_{Sc}/L_{Uc})$ for the two subperiods. I also plot the unconstrained regression line (solid line), and a

TABLE 2.4. Summary Statistics for $(A_{Sc}/A_{Uc})^\rho$

Year	Obs	Min	P10	P50	P90	Max	Corr w/Y
1995	82	0.02	0.11	0.25	0.47	1	0.36
2005	84	0.01	0.07	0.18	0.42	1	0.33

Relative to the United States. Px = xth percentile. Y is income per worker.

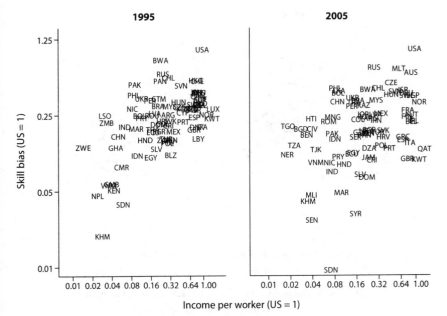

Figure 2.3. Skill-Biased Technology Differences

regression line constrained to have slope equal to −0.66 (denoted "Counterfactual Values").

As predicted by the theory, there is a negative relationship in the data. But the unconstrained regression line is much flatter than the theoretical one of $(\rho - 1)$: the slope is only −0.14 in both periods (standard errors 0.05 and 0.06, respectively). In other words, skill premia do decline with relative skill supply, as predicted by the theory. But they do not decline nearly as fast as they should given an elasticity of substitution of 1.5. This suggests that there is an omitted variable that is positively correlated with both skill premia and relative skill supplies. This variables is $\log(A_{Sc}/A_{Uc})$. In particular, it must be that some technological factor slows down the decline of the relative marginal productivity of skilled labor as the relative supply of skilled labor increases, i.e. there must be a skill bias in technology differences.

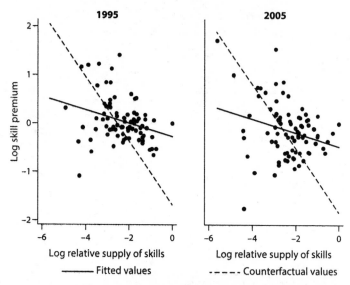

Figure 2.4. Actual and Counterfactual Relationship between Skill Premia and Skill Supply

This discussion also offers us a way of assessing the robustness of the skill bias result to alternative choices of the elasticity of substitution. Clearly the higher the elasticity of substitution, the flatter the predicted relationship between skill premia and skill supplies, the less we need to appeal to skill bias differences in technology to rationalize the data. So how large does the elasticity of substitution need to be to eliminate the skill bias result? Or, in other words, what value of the elasticity of substitution will make the counterfacutal lines in figure 2.4 coincide withe the fitted lines? Since both fitted lines have a slope of −0.14 the answer is (implicitly) given by solving the equation $(1 - \rho) = 0.14$, implying an elasticity of substitution in excess of 7. This is utterly outside all reasonable bounds for the elasticity of substitution estimated in the literature.

The evidence in this chapter is of course the cross-country analogue of time series evidence of skill-biased technical change over the last decades of the 20th century (to which I return in the second part of this book). During that period skill premia failed to decline, indeed they rose, during periods where the relative supply of skills also increased. Labor economists and macroeconomists have argued that increases in the relative efficiency of skilled labor have countered the depressing effect on their relative wages from the increase in their relative supply. The evidence presented here implies that a similar mechanism is required to interpret patters of wage variation across countries.

2.6 Alternative Skill Thresholds

So far I have used "high school completed" as my threshold for "skilled." This choice was dictated by the goal of matching the definition of skilled in Ciccone and Peri (2005), which is my source for the elasticity of substitution. However, other considerations may militate in favor of alternative choices of threshold.

Clearly there is no obvious way to establish a priori which is the best way of splitting workers into the two broad "unskilled" and "skilled" categories. Workers within each of the two sub-aggregates are assumed to be perfect substitutes (though of course with different efficiency units), whereas workers across sub-aggregates are assumed to be imperfect substitutes. Heuristically, differences within groups are "quantitative"—some workers are more productive than others—but differences between groups are "qualitative,"—some workers are fundamentally different. Reality is obviously much more nuanced, and drawing an arbitrary line to classify workers in these two categories is a subjective judgment.

Having said that, one may argue that a definition of "skilled" based on primary schooling completed, rather than secondary completed, may more closely capture a "qualitative" break. This definition roughly separates out the completely illiterate and innumerate from those who can at least read a simple text (e.g. a simple set of instructions or a newspaper article) and perform some basic arithmetic operations. There are many tasks that no number of completely illiterate agents will be able to perform. Beyond the literacy threshold, most increases in education may be seen to have more of an incremental effect on skills, in the sense that most (though admittedly not all) production-relevant tasks that require literacy are accessible to all literate workers— though the less educated will need more time to perform them. Hence the assumption that all workers who are at least literate are perfect substitutes is possibly more defensible than the assumption that the completely illiterate are perfectly substitutable with, say, those with some high school education (but not with college).

At the other end of the spectrum, others may regard the completion of a college education as the major qualitative step in one's accumulation of skills. And of course in empirical work on the United States and other rich countries it is customary to identify the college educated as the skilled in the labor force.

For these reasons, in this section I sketch robustness checks of the skill bias result to these two alternative choices of the skill threshold. The general strategy is identical to the one followed in the rest of this chapter but I do take some shortcuts to economize on computations. Instead of re-estimating the equivalent of equations (2.6) and (2.7) for the alternative definitions of skills, I use the formulas

$$U_c = \sum_{j=1}^{\bar{j}-1} e^{0.10 s_{jc}} l_{jc}$$

$$S_c = \sum_{j=\bar{j}}^{7} e^{0.10\left(s_{jc}-s_{\bar{j}c}\right)} l_{jc}$$

to construct the aggregate supplies of skilled and unskilled labor. In these expressions, $\bar{j}$ is the attainment level that triggers classification into the skilled pool (e.g. $\bar{j} = 3$ when using completed primary), and, recall, s_{jc} are the years of schooling of a worker with attainment j. The coefficient 0.10 is the "consensus" estimate of the Mincerian return in the labor literature. Hence, I am essentially approximating the nonlinear model in equations (2.2)

TABLE 2.5. Summary Statistics for $(A_{Sc}/A_{Uc})^{\rho}$: Robustness to Skill Threshold

Skill Threshold	Year	Obs	Corr w/Y	Implied EOS
Primary	1995	82	0.45	20
	2005	84	0.42	100
Secondary	1995	82	0.41	11
	2005	84	0.36	14
College	1995	82	0.24	20
	2005	84	0.23	14

Y is income per worker. Implied EOS is the elasticity of substitution consistent with skill neutrality.

and (2.3) by a (log-)linear one. In a similar spirit, I approximate the skill premium as in Caselli and Coleman (2006) by

$$\frac{W_{Sc}}{W_{Uc}} = e^{b_c s_{\bar{j}c}},$$

where, recall, b_c is the (country-specific) Mincerian return. I present results using high school completion as a threshold, alongside the two alternative thresholds, to verify that these shortcuts do not lead to excessively different results.

The results for alternative skill thresholds are presented in table 2.5. With all three skill thresholds there is a clear positive association between $(A_{Sc}/A_{Uc})^{\rho}$ and income per worker. The correlation using the "secondary completed" threshold is only marginally larger than using the full procedure to estimate skill premia, so the shortcut described above is probably harmless. These results impose the same elasticity of substitution of 1.5, regardless of the skill threshold. To gauge the robustness of these results to alternative choices of ρ, the last column of the table reports the elasticity of substitution implied by the regression of log skill premia on log relative supply of skills. Recall from the discussion above that the (negative of the) inverse of this coefficient is the elasticity of substitution consistent with no skill bias. It is apparent from the figures in the table that an absence of skill bias can be reconciled with the data on wage premia and relative skill supply only when skilled and unskilled workers are near perfect substitutes.

2.7 Implications of Differences in School Quality

The analysis above assumes that workers with the same educational attainment are comparable across countries. Among other things, this implies that they embody similar amounts of *cognitive skills*. There are many possible reasons to challenge this assumption. Potential sources of systematic differences in cognitive skills include cross-country differences in average health [Weil (2007)], on-the job learning [Lagakos et al. (2012)], school quality [e.g. Hanushek and Woessman (2012)], and parental inputs [De Philippis and Rossi (2016)].

If cross-country differences in cognitive skills are invariant across attainment levels their omission from the analysis has no implication for our estimates of the skill bias. Their inclusion affects the effective supply of skilled and unskilled labor, but not their relative supply. But it is the relative supply of skills that enters the calculation of the skill bias.

One may be concerned, however, that cognitive-skill differences affect skilled workers disproportionately. Take the case of differences in school quality. Clearly omission of these differences will not bias the estimated supply of efficiency units by workers with no schooling, but they will affect the estimated supply of efficiency units by workers with some schooling. Furthermore, even among workers with schooling, it is plausible that the effect of school quality is cumulative. The more time workers spend in school, the larger the impact of school quality on their skill endowment.

Perhaps counterintuitively, if differences in school quality disproportionately affect skilled workers, this may result in an *underestimate* of the skill bias. The reason is the following. Begin by noticing that school quality is, by all possible measures, positively correlated with average educational attainment: high observed L_S/L_U is associated with better quality [e.g. Hanushek and Woessman (2012)]. Therefore, the *effective* relative supply of skills L_S/L_U is underestimated in high observed L_S/L_U countries relative to low L_S/L_U countries. But then, an even higher A_S/A_U is required in these countries to explain the relatively high relative marginal productivity of skills, as captured by the relative wage.

The following example may help consolidate this intuition. Assume only two countries and only two levels of achievement: no schooling and some schooling. Workers with no schooling embody the same amounts of skills in the two countries, but country 2 has better schools, so workers with schooling embody more skills in country 2 than in country 1. In particular, the "true" skill endowment S_c in country c is

$$S_c = (1 + q_c)\tilde{S}_c,$$

for $c = 1, 2$, where $\tilde{S}_c$ measures the share of the labor force with some schooling and $0 < q_1 < q_2$.

While equation (2.1) continues to describe the relationship between "true" relative marginal products, relative efficiencies, and relative skill supplies, if we use the methodology of this chapter (which ignores school quality) we retrieve

$$\left(\frac{\tilde{A}_{Sc}}{\tilde{A}_{Uc}}\right)^{\rho} = \frac{W_{Sc}}{W_{Uc}}\left(\frac{\tilde{S}_c}{U_c}\right)^{1-\rho} = \left(\frac{1}{1+q_c}\right)^{1-\rho}\left(\frac{A_{Sc}}{\tilde{A}_{Uc}}\right)^{\rho}.$$

Hence, the greater the school quality, the more underestimated the relative efficiency with which the country uses skills. Finally, assume that school quality is positively correlated with attainment, i.e. $S_2 > S_1$ (note that $S_c = 1 - U_c$ in the present context). Then, the extent of skill bias is underestimated as we underestimate A_S/A_U more in the country with higher (measured and true) relative skill supply. In other words, the omission of cognitive skills leads to an *underestimate* of the skill bias!

Note that in writing the latest expression we implicitly assumed that school quality does not affect our procedure to back out the skill premium W_S/W_U. In the context of the current example, this is indeed the case.[16] In the case with many achievement groups, using in (2.9) the β_js from a high–school quality country (as I am doing) biases down the wage premium in low–school quality countries. In particular, if the effect of school quality is cumulative, the β_js will be more steeply increasing in j (within each broad skill category) in high–school quality countries than in low–school quality countries. Now a given value of the Mincerian return can result from different combinations of the "within-group" wage gradient and the "between-groups" wage gap. The steeper the within-group wage gradient, the smaller the between-group gap. If in low–school quality countries the within-group gradient is flatter than we impose, then the "true" wage premium is larger than we estimate. Hence, there is a slight (and hard to quantify) element of ambiguity in the general case with school quality: while the variance of L_S/L_U is underestimated (leading to an under-estimate of the skill bias), the variance of the skill premium is also underestimated (leading to an overestimate of the skill bias). The fact that only the former effect is present in the two–schooling levels example leads me to conjecture that it should be considered the dominant effect.

2.8 Implications of Capital-Skill Complementarity

As discussed in the introductory chapter, there are multiple poten-tial CES nestings of the four factors of production I consider in this study. For example, an alternative possibility would have been

$$Y_c = \left\{ (A_{Uc}U_c)^\omega + \left[(A_{Sc}S_c)^\theta + (A_{Kc}K_c)^\theta \right]^{\omega/\theta} \right\}^{1/\omega} . \quad (2.11)$$

---

[16] Equations (2.7) and (2.6) become simply $\log W^i = \log W_U + \varepsilon^i$ for workers with no schooling and $\log W^i = \log W_S + \varepsilon^i$ for workers with schooling, but W_S continues to be the marginal productivity of the average worker with schooling, and hence accurately reflects (countrywide) school quality. Then it is immediate that the analysis of section 2.3 delivers W_S/W_U. It may appear that the assump-tion that ε^i is uncorrelated with s^i is less tenable in the present setting. But this is not the case as q is a country-level variable that is the same for all is.

As emphasized by Krusell, Ohanian, Rios-Rull, and Violante (2000) the potential advantage of this functional form is to allow for a version of capital-skill complementarity. In particular, if $\omega > \theta$, an increase in the supply of physical capital increases the skill premium. One may thus wonder whether the finding that A_S/A_U is higher in high-income countries is driven by not having taken into account this capital-skill complementarity effect.

In Caselli and Coleman (2006) we used (2.11) to perform an exercise similar to that performed in this chapter. In particular, we backed out not only A_u and A_s, but also A_k. This required complementing (2.1) with an additional equation, based on an international no-arbitrage condition on the return to capital. We experimented with a wide range of values for ω and for θ, finding overwhelming evidence of nonneutrality and skill bias. This was also the case when using the Krusell et al. (2000) estimates of these parameters. Since the Krusell et al. parameters imply capital-skill complementarity, the skill bias result is clearly not driven by the failure to account for capital-skill complementarity.

3

NATURAL AND REPRODUCIBLE CAPITAL

3.1 Estimating the Reproducible-Capital Bias

This chapter seeks to identify possible factor biases in the way different countries use reproducible and natural capital. The focus is thus equation (1.11), which for convenience I repeat here:

$$\tilde{K}_c = \left[(N_c)^{\eta} + \left(\frac{A_{Mc}}{A_{Nc}} M_c \right)^{\eta} \right]^{1/\eta}. \qquad (3.1)$$

Recall that $\tilde{K}_c$ is a bundle of capital goods which, combined with labor, is used to produce GDP, N_c is natural capital, and M_c is reproducible capital. The goal of the chapter is to characterize how the factor bias $(A_{Mc}/A_{Nc})^{\eta}$ varies across countries, and in particular how it varies with income per worker Y_c.

Let's begin as in chapter 2 by writing the ratio of the marginal products of the two factors of production. Under perfectly competitive markets for reproducible and natural capital the system (1.10)–(1.12) implies

$$\frac{MPM_c}{MPN_c} = \left(\frac{A_{Mc}}{A_{Nc}} \right)^{\eta} \left(\frac{M_c}{N_c} \right)^{\eta-1}. \qquad (3.2)$$

As before, then, backing out the factor bias $(A_{Mc}/A_{Nc})^{\eta}$ requires three ingredients: relative supply M_c/N_c; relative marginal products MPM_c/MPN_c; and elasticity of substitution η. The last two are considerably more challenging than in the case of skilled and unskilled labor, as I explain below.

3.2 Estimating the Relative Supply of Reproducible Capital

World Bank (2011) presents cross-sectional estimates of the *total* capital stock, as well as its components, for various years.

The total capital stock includes reproducible capital, but also land, timber, mineral deposits, and other items that are not included in standard national account–based data sets.

For reproducible capital, the World Bank uses a standard perpetual-inventory calculation based on historical investment series. For natural capital, the basic strategy begins with estimates of the rental flows accruing from different types of natural capital, which are then capitalized using fixed discount rates. In most cases, the measure of rents is based on the value of output from that form of capital in a given year. For subsoil resources, the World Bank also needs to estimate the future growth of rents and a time horizon to depletion. For forest products, rents are estimated as the value of timber produced (at local market prices where possible) minus an estimate of the cost of production. Adjustments are made for sustainability based on the volume of production and total amount of usable timberland. The rents from other forest resources are estimated as fixed value per acre for all nontimber forest. Rents from cropland are estimated as the value of agricultural output minus production costs. Production costs are taken to be a fixed percentage of output, where that percentage varies by crop. Pasture land is similarly valued. Protected areas are valued as if they had the same per-hectare output as crop and pasture land, based on an opportunity cost argument. Because of data limitations, no good estimates of the value of urban land are available. A very crude estimate values urban land at 24% of the value of reproducible capital.[1]

---

[1] See Caselli and Feyrer (2007) for further discussions as well as for checks on the reliability of these data (though the current data pertain to a revised version of the data set).

TABLE 3.1. Summary Statistics for M_c/N_c

Year	Obs	Min	P10	P50	P90	Max	Corr w/Y
1995	120	0.01	0.03	0.14	1.37	5.68	0.52
2005	144	0.01	0.04	0.11	1.45	8.96	0.50

Relative to the United States. Px = xth percentile. Y is income per worker.

In the calculations below, I map the notion of reproducible capital (natural capital) to the variable *producedplusurban* (*natcap*) in the World Bank's data set. The data is available for 124 countries in 1995 and 151 countries in 2005. However four very large outliers appear in the distribution of M/N in both years, so I drop them from the rest of this chapter's analysis.[2] Furthermore, I drop countries for which I do not have income per worker from the Penn World tables.[3] Table 3.1 reports summary statistics and figure 3.1 plots the relative supply of reproducible capital against relative income per worker in the remaining sample. Similar to the case of the relative supply of skills, there is considerable dispersion in the relative supply of reproducible capital, with richer countries having a larger relative endowment of reproducible capital.

3.3 Relative Marginal Productivities

Recall that in estimating the skill bias we relied heavily on the relationship between marginal productivities and wages, and we used the fact that wages are observable. Here we correspondingly exploit the relationship between marginal products of the capital inputs and the respective rental rates. However, we need to get around the fact that rental rates are not directly observable.

Caselli and Feyrer (2007) point out that the marginal productivity of an input can be estimated from data on the share of output

---

[2] The outliers are Saint Lucia, Hong Kong, Macao, and Singapore.

[3] Saint Kitts and Nevis in both years, and Dominica, Seychelles, and Grenada in 2005.

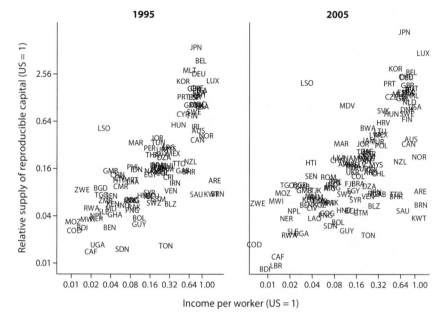

Figure 3.1. M_c/N_c against Y_c

going to that input, the amount of output, and the amount of the input. In particular, under perfect competition,

$$MPX_c = SHX_c \frac{Y_c}{X_c}, \qquad (3.3)$$

where MPX is the marginal productivity of factor X, SHX is the share of output accruing to factor X, and X is also used to denote the quantity of X. This equation holds because the right-hand side is nothing but the rental rate for factor X. Applied to the issue at hand, this formula is useful if we can recover estimates of the share in GDP of natural and reproducible capital—as we obviously already have GDP and the quantities of the two capital types.

The Penn World tables report estimates of the labor share in GDP. These estimates should be treated with considerable caution, as I discuss in greater detail in the next chapter. Nevertheless, I use them here to compute the *total* capital share, as 1 minus the labor share. The question now is how to apportion the total capital share between total and natural capital, so that I can use equation (3.3) to back out the marginal products of the two types of capital.

Recently, Monge-Naranjo et al. (2015) noticed that the same World Bank data set that I used for M and N also reports estimates of total rents paid out for most of the components of the natural capital stocks. For the few for which rents are not reported, the data set documentation and Monge-Naranjo et al. suggests relatively straightforward methods to fill in the gaps. Using these data and following these methods we can directly measure the share of natural capital in output as

$$SHN_c = \frac{\text{Rents to } N_c}{Y_c}.$$

Then, the share accruing to reproducible capital is simply

$$SHM_c = SHK_c - SHN_c,$$

where SHK_c is the total capital share.

A disadvantage of this approach is that, with the rents to natural capital and GDP coming from different data sources, differences in methodology, measurement error, or other anomalies imply that SHN is not always less than 1. Furthermore, SHM is not always greater than 0.[4]

---

[4] Caselli and Feyrer (2007) originally proposed "sharing" the total capital share in proportion to the shares of the two types of capital in total capital, or

$$SHN_c = \frac{N_c}{M_c + N_c} SHK_c.$$

But this procedure assumes that rental rates, and hence marginal products, are equalized (within countries) between natural and reproducible capital. In turn, this requires not only that rates of returns to investing in the two types of capital are equalized, but also that the capital gain component of the returns to capital is very small relative to the rental rate component. These are stringent assumptions but, more to the point here, under this method there would be no variation whatsoever across countries in MPM/MPN.

TABLE 3.2. Summary Statistics for MPM_c/MPN_c

Year	Obs	Min	P10	P50	P90	Max	Corr w/Y	Corr w/Y (trim)
1995	86	−1.70	0.28	1.60	7.85	442.10	0.13	−0.43
2005	106	−6.13	0.37	1.56	8.99	204.26	−0.10	−0.34

Relative to the United States. Px = xth percentile. Y is income per worker. "trim" refers to a subsample where the bottom and top decile of the MPM/MPN distribution have been dropped

With these caveats, summary statistics for relative marginal productivities MPM/MPN (relative to the United States) for the countries for which data is available are reported in table 3.2. Clearly with some of the estimates being negative and some being outrageously high it is hard to be terribly confident in their validity. The raw correlations with GDP are very low. However, when dropping the estimates of MPM/MPN in the top and bottom decile the correlation appears to be stronger and negative. Remembering that high-income countries have higher ratios of reproducible capital relative to natural capital, if taken at face value the correlation suggests that high relative supplies of reproducible capital are associated with low relative products of reproducible capital. The corresponding scatterplots, for the trimmed sample only, are in figure 3.2.

3.4 Inferring the Bias toward Reproducible Capital

We have seen from (3.2) that

$$\left(\frac{A_{Mc}}{A_{Nc}}\right)^{\eta} = \frac{MPM_c}{MPN_c}\left(\frac{M_c}{N_c}\right)^{1-\eta}. \qquad (3.4)$$

Recall from figure 3.1 and table 3.1 that there is massive variation in M/N across countries, with rich countries having systematically higher M/N. The evidence on MPM/MPN is much more tentative, because questions of data quality hang even heavier than usual. Nevertheless, the tentative evidence points

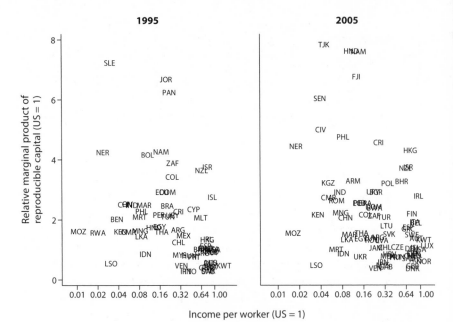

Figure 3.2. MPM_c/MPN_c against Y_c

to a negative relation between income and relative marginal products. Taking this evidence at face value, we are left with ambiguous predictions on whether rich countries use reproducible capital relatively efficiently compared to poor countries. This is entirely similar to the situation we encountered with regard to skill bias, where relative wages were higher in poor countries and relative supplies were higher in rich countries.

To resolve the ambiguity, we would need of course an estimate of the elasticity of substitution $1/(1 - \eta)$. Clearly the higher the value of the elasticity of substitution (i.e. the higher η), the higher the relative impact of MPM/MPN on our estimate of the factor bias in technology. In particular, experimenting with different values of η in the trimmed sample suggests that there is a bias toward natural capital (in both periods) if the elasticity of substitution exceeds a value in the neighborhood of 3. If the elasticity of substitution is below this value, there is a bias in favor of reproducible capital: $(A_{Mc}/A_{Nc})^{\eta}$ is higher in rich countries.

Unfortunately, to my knowledge no attempts have been made to estimate this elasticity. Introspection, however, suggests that a value in excess of 3 is unrealistic. For some industries the elasticity of substitution obviously must be much smaller: agriculture requires land and tools; mineral extraction requires mineral deposits and equipment suitable for extraction. But even more generally, it seems implausible that reproducible and natural capital are more than twice as substitutable as skilled and unskilled labor (recall that the elasticity of substitution between skilled and unskilled labor is 1.5). I conclude that, on balance, technology differences across countries are biased toward reproducible capital relative to natural capital.[5]

---

[5] We would have reached the same conclusion had we followed Caselli and Feyrer's approach to apportioning the capital share between reproducible and natural capital. This is for the simple reason that, as discussed in the previous footnote, under their assumptions we have $MPM_c = MPN_c$, so only variation in M_c/N_c exists in equation (3.4).

4

CAPITAL AND LABOR

4.1 Inferring Augmentation Coefficients for Labor and Capital

The previous two chapters looked *within* the labor input and *within* the capital input to detect biases toward skilled labor and, respectively, reproducible capital. In this chapter we go "up one level" and look at the efficiency with which the aggregate labor input and, respectively, the aggregate capital input are used in production. In order to do so, we work with the representation

$$Y_c = [(\tilde{A}_{Kc}\tilde{K}_c)^\sigma + (\tilde{A}_{Lc}\tilde{L}_c)^\sigma]^{1/\sigma}, \tag{4.1}$$

and we seek to tease out patterns of variation in $\tilde{A}_{Kc}$ and $\tilde{A}_{Lc}$.[1] These coefficients operate as augmentation coefficients for "natural capital equivalents" $\tilde{K}$, i.e. the capital input expressed in efficiency units of natural capital, and "unskilled-labor equivalents" $\tilde{L}$, or the labor input in efficiency units of unskilled labor.

---

[1] Recall from section 1.6 that the higher-level production function can be written as
$$Y_c = [(A_{Kc}A_{Nc}\tilde{K}_c)^\sigma + (A_{Lc}A_{Uc}\tilde{L}_c)^\sigma]^{1/\sigma},$$
where
$$\tilde{K}_c = \left[(N_c)^\eta + \left(\frac{A_{Mc}}{A_{Nc}}M_c\right)^\eta\right]^{1/\eta},$$
$$\tilde{L}_c = \left[(U_c)^\rho + \left(\frac{A_{Sc}}{A_{Uc}}S_c\right)^\rho\right]^{1/\rho},$$
so that $\tilde{A}_{Kc} = A_{Kc}A_{Nc}$ and $\tilde{A}_{Lc} = A_{Lc}A_{Uc}$.

We appeal as usual to the implications of (approximately) perfect factor markets to derive, from the production function above, the identity between factor prices and marginal productivities:

$$\tilde{W}_c = \left(\tilde{A}_{Lc} \right)^{\sigma} \left(\frac{Y_c}{\tilde{L}_c} \right)^{1-\sigma},$$

$$\tilde{R}_c = \left(\tilde{A}_{Kc} \right)^{\sigma} \left(\frac{Y_c}{\tilde{K}_c} \right)^{1-\sigma},$$

where $\tilde{W}_c$ ($\tilde{R}_c$) are the wage (rental rate) per equivalent unit of unskilled labor (natural capital). These expressions can be inverted and rearranged to yield

$$\tilde{A}_{Lc} = \left(\frac{\tilde{W}_c \tilde{L}_c}{Y_c} \right)^{1/\sigma} \frac{Y_c}{\tilde{L}_c}, \tag{4.2}$$

$$\tilde{A}_{Kc} = \left(\frac{\tilde{R}_c \tilde{K}_c}{Y_c} \right)^{1/\sigma} \frac{Y_c}{\tilde{K}_c}. \tag{4.3}$$

The ratio outside the parentheses is a measure of labor (capital) productivity.[2] It is highly intuitive that high labor (capital) productivity is a symptom of a high level of efficiency in using the labor (capital) input. The expressions in parentheses are immediately recognized as the labor and, respectively, capital shares in income. A high level of efficiency in using labor increases the effective supply of labor, and it will tend to increase the wage bill if labor and capital are good substitutes ($\sigma > 0$) and reduce it if they are poor substitutes ($\sigma < 0$).

Equations (4.2) and (4.3) tell us how we can make inferences on $\tilde{A}_{Lc}$ and $\tilde{A}_{Kc}$ from observables. In particular, we need measures of labor and capital inputs $\tilde{L}_c$ and $\tilde{K}_c$; a measure of per-worker income, Y_c; the labor and capital shares in GDP; and an estimate of the elasticity of substitution between labor and capital, $1/(1-\sigma)$.

---

[2] Here and in the rest of the chapter I use the phrase "labor productivity" in the sense of $Y/\tilde{L}$, or output per unskilled-labor equivalent. To refer to the more usual notion of labor productivity I will continue to use the phrase "income per worker."

The measurement of $\tilde{L}$ and $\tilde{K}$ was essentially the topic of chapters 2 and 3. In those chapters we estimated the relative efficiency of skilled and unskilled labor and the relative efficiency of reproducible and natural capital. With these relative efficiencies at hand one can construct the aggregates $\tilde{L}$ and $\tilde{K}$.[3]

Now admittedly we were more successful in pinning down $\tilde{L}$ than $\tilde{K}$, particularly because an exact estimate of $\tilde{K}$ requires an as yet unavailable estimate of the elasticity of substitution between natural and reproducible capital. Hence, while in this chapter I can use directly the results from chapter 2 to measure $\tilde{L}$, I must take a shortcut to measure $\tilde{K}$. In particular, I revert to the benchmark approach of treating natural and reproducible capital as perfect substitutes. Under this benchmark, we can simply measure $\tilde{K}$, from the World Bank's data used in chapter 3, as the sum of reproducible and physical capital. I refer to this sum as *total capital.*

As in the previous chapters I measure per-worker income from the Penn World tables. Note that it is precisely the fact that we know the value of the aggregate to the left-hand side of (4.1) which allows us to back out the absolute values of the As. In the analysis of previous chapters we did not observe the value of the aggregate on the left-hand side, so all we could do was take the ratio of (the equivalents of) (4.2) and (4.3) and back out the ratio of the As.[4]

Measuring the capital share is another challenge to the implementation of the approach described above. Traditionally, the capital share is measured from the national accounts as a residual after employee compensation has been taken out from GDP. With this method, the capital share is generally found to be higher in poor countries than in rich countries. However, Gollin (2002) has criticized the construction of the traditional estimates of the capital share and has provided revised estimates that—among

---

[3] See footnote 1 in this chapter.

[4] Indeed, one could solve the system constituted by one of (4.2) and (4.3) and equation (4.1). The result would be identical. The properties of the constant returns to scale production function, combined with national-account identities, imply that from any two of these equations the third follows.

other things—attempt to include the labor component of self-employment income in the labor share. These estimates show essentially no systematic pattern of cross-country variation in capital shares. This has been confirmed by Bernanke and Gurkaynak (2001), who extended Gollin's contribution. Unfortunately, these estimates are now quite out of date; Gollin reports the 1980–1995 average of his calculations, and Bernanke and Gurkaynak stop at 1990. It would thus be pretty heroic to use these figures in combination with our 1995, and—even more so—2005 values for $\tilde{L}$, $\tilde{K}$, and Y. As we saw in the previous chapter, the Penn World tables provide recent estimates of the labor share. These estimates do attempt to correct for some of the biases in the national-account estimates, but the attempt appears significantly cruder than Gollin's. Nevertheless, I will look at them shortly for a "reality check" on the argument I am about to present.

Another problem is that there is no consensus on the value of the elasticity of substitution $1/(1-\sigma)$. Hamermesh (1986) provides an exhaustive survey, featuring firm-, industry-, and country-level studies, both cross-sectional and time series. Unfortunately, he reports a dismayingly wide range of estimates, both greater and less than 1. However, a majority of the estimates tend to be less than 1. Antràs (2004) is a relatively recent example.[5]

---

[5] As pointed out by Antràs (2004), the nonneutrality approach we follow here implies an intrinsic pitfall in attempting to identifying this parameter. Specifically, many empirical investigations of the elasticity of substitution implicitly assume no variation across observations in the relative efficiency of labor and capital. If $\tilde{A}_K$ and $\tilde{A}_L$ vary across observations, then the effective input $\tilde{A}_K K$ and $\tilde{A}_L L$ will be mismeasured, perhaps wildly. I believe this may indeed be the reason why estimates of σ are so unstable. I think this point is implicit in the analysis of Diamond, McFadden, and Rodriguez (1978). If the induced measurement error is random, it seems the bias in the estimate of σ should be upward. Intuitively, observations with very different input combinations will appear to have similar output levels, something that is consistent with a high elasticity of substitution. However, if the $\tilde{A}$s vary systematically, the bias could also be downward. Suppose, for example, that $\tilde{A}_x$ and x are positively correlated across observations. Then the data will tend to understate the true variation in effective input, so that less substitutability will appear to be required to explain the observed variation in output.

Since published estimates of σ are neither stable nor reliable, one could perhaps turn to theoretical considerations. There is of course a tradition of arguing that long-run elasticities are higher than short-run, and macroeconomic higher than microeconomic. Ventura (1997) is a particularly convincing example. For our purposes it clearly seems appropriate to focus on a long-run, aggregate interpretation of the elasticity. However, it is not clear that these arguments put a lower bound on $1/(1 - \sigma)$: even accepting that it is higher than a microeconomic, short-run elasticity does not necessarily imply that it is, say, greater than 1.

Combined, the paucity of recent estimates of the capital and labor shares, and the lack of consensus on the value of the elasticity of substitution, make it difficult to pin down values for the first of the two multiplicative terms in equations (4.2) and (4.3). On the other hand, there is no particular reason to suspect that Gollin's headline result of a lack of systematic correlation between labor shares and income per worker would no longer hold in more recent times. As a reality check on this hypothesis, figure 4.1 plots the labor share in the Penn World tables against GDP per worker. The corresponding correlation coefficients are -0.04 and -0.01. Clearly these data give no indication that Gollin's original finding is no longer valid for the more recent periods.

If we are willing to conclude that there is still little or no correlation between labor shares and income per worker, then neither the lack of precise up-to-date data nor the uncertainty surrounding σ need prevent us from making inference on $\tilde{A}_L$ and $\tilde{A}_K$. In particular, regardless of the value of σ, $\tilde{A}_L$ and $\tilde{A}_K$ will be roughly proportional to $Y/\tilde{L}$ and $Y/\tilde{K}$, respectively. Or, more precisely, the correlation between $\tilde{A}_L$ ($\tilde{A}_K$) and income per worker will be approximately equal to the correlation between $Y/\tilde{L}$ ($Y/\tilde{K}$) and income per worker.

Table 4.1 and figure 4.2 present summary statistics and scatterplots of $Y/\tilde{L}$, which proxies for $\tilde{A}_L$ (up to a multiplicative term uncorrelated with income per worker). In both periods the cross-country pattern of technology differences is *labor augmenting*. The higher is income per worker, the higher the efficiency

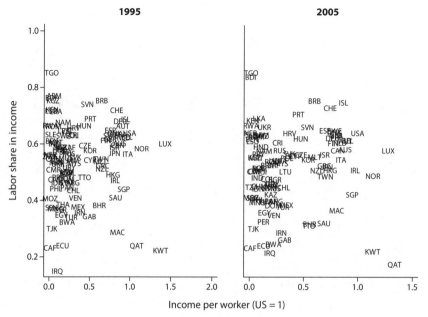

Figure 4.1. Labor Share against Income

TABLE 4.1. Summary Statistics for $\tilde{A}_{Lc}$

Year	Obs	Min	P10	P50	P90	Max	Corr w/Y
1995	82	0.88	1.79	10.22	26.67	63.52	0.61
2005	84	0.86	2.65	14.89	58.17	221.63	0.55

Relative to the United States. Px = xth percentile. Y is income per worker.

with which labor is used. This qualitative pattern is pretty un-surprising: we are using "income per unskilled worker equivalent" as a proxy for the augmentation coefficient, and it might have been expected that this would be fairly highly correlated with income per worker—though of course differences across countries in schooling could drive a large wedge between the two.

What may seem more surprising is that the United States stands in sharp contrast to the overall pattern: it has one of the lowest

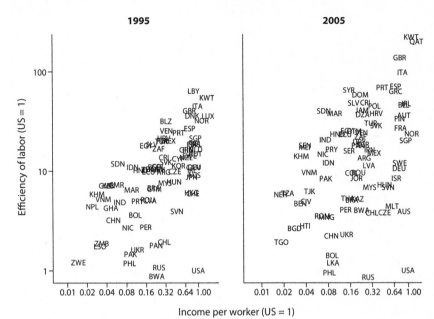

Figure 4.2. $\tilde{A}_{Lc}$ against Y_c

labor productivities in the sample. To understand this seemingly odd feature it is crucial to remember that $\tilde{L}_c$ is quality adjusted. First, it adjusts for the skill composition of the labor force. Second, it weighs skilled workers by their relative efficiency compared to unskilled workers. On both counts, the United States has a massive advantage on all other countries (cfr. figures 2.1 and 2.3, paying close attention on the axes labels), resulting in an estimated value of $\tilde{L}_{USA}$ that is absolutely vast compared to any other country in the sample, both in absolute terms and, more important, relative to the number of workers.

Summary statistics and scatterplots of $Y/\tilde{K}$—a proxy for $\tilde{A}_K$—are reported in table 4.2 and figure 4.3. Here the surprising result is that capital productivity is *decreasing* in per capita income: richer countries use capital *less* efficiently. I defer an interpretation of this result to part II of the book.

TABLE 4.2. Summary Statistics for $\tilde{A}_{Kc}$

Year	Obs	Min	P10	P50	P90	Max	Corr w/Y
1995	120	0.24	0.50	0.91	1.65	2.55	−0.13
2005	144	0.25	0.55	0.99	1.67	2.45	−0.26

Relative to the United States. Px = xth percentile. Y is income per worker.

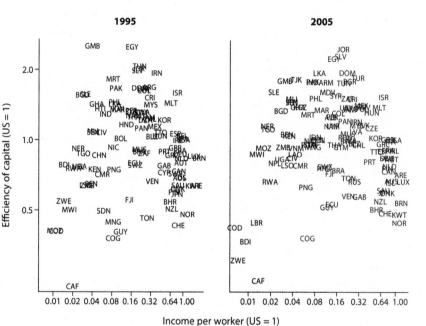

Figure 4.3. $\tilde{A}_{Kc}$ against Y_c

It is worth remarking that, despite the surprising negative correlation of capital efficiency with income, the results in this section are easily reconciled with the common development-accounting wisdom that overall total factor productivity (TFP) is higher in high-income countries (as discussed in the introductory chapter). Consider the standard Cobb-Douglas specification $Y_c = \tilde{A}_c \tilde{K}_c^{\alpha} \tilde{L}_c^{1-\alpha}$. One way of writing TFP is

$$\tilde{A}_c = \left(\frac{Y_c}{\tilde{K}_c}\right)^\alpha \left(\frac{Y_c}{\tilde{L}_c}\right)^{1-\alpha}.$$

Hence, the conclusion that rich countries have higher TFP is based on the fact that the increasing pattern of $(Y_c/\tilde{L}_c)^{1-\alpha}$ more than compensates for the decreasing pattern in $(Y_c/\tilde{K}_c)^\alpha$.

Obviously these patterns in the absolute values of $\tilde{A}_L$ and $\tilde{A}_K$ also mean that *relative* labor efficiency $\tilde{A}_L/\tilde{A}_K$ is increasing in per capita income. Whether this means that technology differences are biased toward labor or toward capital depends on the elasticity of substitution $1/(1 - \sigma)$. The vast majority of estimates of the elasticity of substitution between capital and labor is below 1, implying $\sigma < 0$. With $\sigma < 0$, $(\tilde{A}_L/\tilde{A}_K)^\sigma$ is decreasing in per capita income, so technology differences are *biased toward capital*.

In the next chapter I will present a theoretical framework capable of rationalizing these findings, including the negative relationship between Y_c and $\tilde{A}_{Kc}$, as well as those from the previous two chapters. First, however, I devote the rest of this chapter to probing the robustness of the benchmark results above.

4.2 Variable Capital Shares

Besides being somewhat out of date, the Gollin and Bernanke and Gurkaynak data sets are also somewhat small, and developed economies are overrepresented. Furthermore, many untested assumptions have been used to develop these estimates. Hence, the conclusion that capital shares are not systematically related to labor productivity is not ironclad. What would it take then to reverse the startling result that poor countries are more efficient users of capital?

If factor shares vary systematically with per-worker income, then it becomes critical to know what is the elasticity of substitution $1/(1 - \sigma)$. Suppose that the capital share is higher in rich countries. If $\sigma > 0$ (i.e. capital and human capital are good substitutes relative to the Cobb-Douglas case), then $\tilde{A}_K$ may conceivably

TABLE 4.3. Predicted Correlations between $\tilde{A}_K$ and Y_c, and $\tilde{A}_L$ and Y_c

	$\mathrm{Corr}(\tilde{R}_c\tilde{K}_c/Y_c, Y_c) > 0$	$\mathrm{Corr}(\tilde{R}_c\tilde{K}_c/Y_c, Y_c) = 0$	$\mathrm{Corr}(\tilde{R}_c\tilde{K}_c/Y_c, Y_c) < 0$
$\sigma > 0$	?,?	$-,+$	$-,+$
$\sigma < 0$	$-,+$	$-,+$	?,?

become increasing in income [if $(\tilde{R}_c\tilde{K}_c/Y_c)^{1/\sigma}$ grows "faster" than $Y_c/\tilde{K}_c$ falls]. In this case, however, since the labor share is 1 minus the capital share, the result on $\tilde{A}_L$ could also possibly be overturned. If $\sigma < 0$, the results from the constant-share case would be reinforced. Symmetrically, if the capital share is decreasing in income, the negative (positive) correlation between $\tilde{A}_K$ ($\tilde{A}_L$) and Y_c would be reinforced for $\sigma > 0$, and weakened (and possibly overturned) if $\sigma < 0$. These observations are summarized in table 4.3. Each cell of the table lists the predicted sign (positive, negative, or ambiguous) for the correlation between $\tilde{A}_K$ and Y (first term) and between $\tilde{A}_L$ and Y (second term), conditional on the observed patterns of $Y_c/\tilde{K}_c$ and $Y_c/\tilde{L}_c$, under various assumptions on σ and on the correlation between $\tilde{R}_c\tilde{K}_c/Y_c$ and Y_c.

The intuition for the way observed factor shares modify our predictions on cross-country efficiency patterns is simple. If $\sigma > 0$, the two factors are good substitutes. Because the two factors are good substitutes, it makes sense to try to increase the usage of the most efficient factor. Hence, when $\sigma > 0$ demand will concentrate on the factor with high efficiency, leading to a high share in income for this factor. Conversely, then, with $\sigma > 0$, when we observe a high income share for factor x we can infer that this factor is efficient. On the other hand, if $\sigma < 0$, the two factors are poor substitutes. In this case, allocative efficiency calls for boosting the overall efficiency units provided by the low-efficiency factor. This increases the income share of this factor. Hence, with $\sigma < 0$, a high income share for factor x signals that this factor is used inefficiently.

In sum, skepticism about the greater capital efficiency of poor countries is reasonable only if one believes that there is a strong positive correlation between the capital share and income *and* $\sigma > 0$ (or the elasticity of substitution is greater than 1); *or* if one believes that there is a strong negative correlation between the capital share and Y_c *and* $\sigma < 0$. In all other cases the result from the previous section is robust.

4.3 A Broader Measure of Labor Inputs

So far in this chapter I have used a measure of $\tilde{L}_c$, based on the analysis in chapter 2, which takes into account the distribution of the labor force among different educational achievement categories, as well as the skill bias in technology. However, the literature has identified at least two potential additional factors that affect how the effective (quality-adjusted) labor input varies across countries. One of these is the health status of the labor force. The other is the *cognitive skills* embodied in workers (holding constant the quantity of schooling).

Weil (2007) pointed out that healthy workers are more productive, and that overall rates of morbidity from various illnesses vary substantially across countries. It follows that health may be an important consideration in constructing cross-country comparisons of quality-adjusted labor inputs.

To account for differences in health, I follow Weil and augment the previously constructed measure of $\tilde{L}_c$ by the factor $\exp(\beta_H H_c)$, where H_c is the *adult survival rate* and β_H is a parameter that maps variation in adult survival into (proportional) variation in human capital. The adult survival rate is a statistic computed from age-specific mortality rates at a point in time. It can be interpreted as the probability of reaching the age of 60, conditional on having reached the age of 15, *at current rates of age-specific mortality*. Since most mortality before age 60 is due to illness, the adult survival rate is a reasonably good proxy for the overall health status of the population at a given time. Relative to more direct measures of health, the advantage of the adult survival rate is that it is available for a large cross section of countries.[6]

The calibration of β_H is also taken from Weil. He uses times series evidence from a few countries to establish a mapping between changes in survival rates and changes in height. This he then combines with micro evidence on the relationship between height and wages, to arrive at an overall mapping between differences in survival rates and differences in wages/human capital. The resulting value of β_H is 0.65. This means that if the survival rate goes from 0 to 1, human capital increases by 65 percent. To put this in context, if the Mincerian return is 0.10, one extra year of schooling generates roughly the same increase in human capital as a 15 percentage-point increase in the adult survival rate.

Recent research by, e.g. Hanushek and Kimko (2000), Gundlach et al. (2001), and Hanushek and Woessman (e.g. 2012) has emphasized that there is substantial cross-country variation in the scores of standardized tests administered to children in given school grades. Various interpretations of the fact that children in the same school year perform very differently on similar tests are possible. Clearly one possibility is that the differences in performance reflect differences in the quality of the education imparted. But De Philippis and Rossi (2016) show that differences in test scores also derive from systematic cross-country differences in parental inputs and home environments.[7]

Regardless of the interpretation, test score results alert us to differences in human capital that should be accounted for in

---

[6] I construct the adult survival rate from the World Development Indicators. Specifically, this is the weighted average of male and female survival rates, weighted by the male and female share in the population. For the "1995" cross section I use data from 1998, as there are too many missing values in 1995.

[7] Yet another possibility is that differences in test scores are at least in part due to differences in children's health. As such they would already be accounted for by the adult-survival correction. However, as we will see below, test scores are drawn from a subsample of countries with relatively high incomes and health, so it is likely that in this subsample health is not a major determinant of test scores.

building a quality-adjusted measure of labor input. Ideally, one would have access to a measure of average cognitive ability in the working-age population. Hanushek and Zhang (2009) report estimates of one such test, the International Adult Literacy Survey (IALS), but it features only a dozen countries.

As a fallback, I rely on internationally comparable test scores taken by school-age children. In particular, I will use scores from a science test administered in 2009 to 15 year olds by PISA (Program for International Student Assessment). Using the notation T_c for the average test score in country c, I then further augment $\tilde{L}_c$ by the factor $\exp(\beta_T T_c)$, where β_T is a parameter mapping changes in test scores into differences in human capital.

There are in principle several other tests (by subject matter, year of testing, and organization testing) that could be used alternatively to, or combination with, the 2009 PISA science test. However, there would be only modest gains in country coverage by using or combining with other years (the PISA tests of 2009 have the greatest participation). Focusing on one test bypasses potentially thorny issues of aggregation across years, subject, and method of administration. Cross-country correlations in test results are very high anyway, and very stable over time. Data on test score results are from the World Bank's Education Statistics.

Measuring cognitive skills by the above-described test scores is clearly very unsatisfactory, as in most cases the tests reflect the cognitive skills of individuals who had not joined the labor force as of 2005, much less those of the average worker. Implicitly, then, we are interpreting test-score differences in current children as proxies for test-score differences in current workers. If different countries experienced different trends in cognitive skills of children since 1984, this assumption is problematic.

The 2009 PISA science tests are reported on a scale from 0 to 1000, and they are normalized so that the average score *among OECD countries* (i.e. among all pupils taking the test in this set of countries) is (approximately) 500 and the standard deviation is

TABLE 4.4. Summary Statistics for $\tilde{A}_{Lc}$: Robustness to Measurement of Labor

Labor Measure	Year	Obs	Corr w/Y
Schooling only	1995	82	0.61
($\tilde{L}_c$)	2005	84	0.55
Schooling and health	1995	82	0.59
($\tilde{L}_c e^{\beta_H H_c}$)	2005	84	0.54
Schooling, health, and tests	1995	47	0.42
($\tilde{L}_c e^{\beta_H H_c + \beta_T T_c}$)	2005	42	0.50

Y is income per worker.

(approximately) 100.[8] For the calibration of β_T I follow Hanushek and Woessmann (2012), who advocate a value of 0.002.

In table 4.4 I report the correlation with income per worker of the labor efficiencies implied by alternative measures of the labor input. First I reproduce the result using schooling only, from the previous subsection. Next I add Weil's health correction. Finally I further add the correction for cognitive skills based on Hanushek and Woessman. The latter addition causes a considerable drop in sample size, due to limits to the availability of test-result data. The result that richer countries use labor relatively more efficiently seems very robust.

4.4 Imperfect Substitution between Reproducible and Natural Capital

I now turn to the surprising result that richer countries use capital *less* efficiently than poorer ones. In the benchmark calculation

---

[8] I say approximately in parenthesis because the normalization was applied to the 2006 wave of the test. The 2009 test was graded to be comparable to the 2006 test. Hence, the 2009 mean (standard deviation) drifted somewhat away from 500 (100)—though not by much. The PISA math and reading tests were normalized in 2000 and 2003, respectively, so their mean and standard deviation have drifted away slightly more from the initial benchmark. This is one reason why I use the science test for my baseline calculations.

I simply measured $\tilde{K}$ by the World Bank's aggregate of natural and reproducible capital. If these two components of the capital stock are imperfect susbstitutes, this is a potentially biased measure, particularly in light of the vast cross-country variation in the composition of the aggregate capital stock documented in chapter 3. In this section I look at how my headline result varies for alternative values of the elasticity of substitution $1/(1 - \eta)$.

Plugging (3.4) into (3.1) we can rewrite the capital stock in units of natural capital as

$$\tilde{K}_c = N_c \left(1 + \frac{P_{Mc}M_C}{P_{Nc}N_c}\right)^{1/\eta}. \tag{4.4}$$

I observe N_c directly from my World Bank data on natural capital (up to a multiplicative constant). Furthermore, if I continue with the assumption from chapter 3 that the relative prices of natural and reproducible capital are fairly similar across countries, I can also proxy the ratio $P_{Mc}M_C/P_{Nc}N_c$ with the ratio M_c/N_c from the same data set. Then, for each choice of η I can compute $\tilde{K}_c$ and hence $Y_c/\tilde{K}_c$ as an alternative proxy for $\tilde{A}_{Kc}$.

We already know that M/N is positively correlated with income per worker (table 3.1, figure 3.1). It turns out that N is also higher in countries with higher income per worker. Indeed the correlation coefficients of N_c and $1 + P_M M_C/P_N N_c$ with Y_c are both on the order of 0.5. When $\eta = 1$ this leads to the results we saw in the benchmark case: rich countries have much more capital than poor countries and the productivity of capital is smaller in poor countries. When $0 < \eta < 1$ differences in $P_M M_C/P_N N_c$ get amplified (for a given N_c), so the result from the benchmark case becomes even stronger. However, when $\eta < 0$ (i.e. the elasticity of substitution between natural and reproducible capital is less than 1) the correlation between the second term in (4.4) and Y_c obviously becomes negative, so the overall correlation between $\tilde{K}_c$ and Y_c becomes much weaker. In particular, $\tilde{K}_c$ now rises less fast than income, so that capital productivity $Y_c/\tilde{K}_c$ becomes increasing in Y_c. Table 4.5 illustrates these mechanisms for various possible

TABLE 4.5. Summary Statistics for
$\tilde{A}_{Kc}$: Robustness to Elasticity of
Substitution between M_c and N_c

$1/(1 - \eta)$	Year	Obs	Corr w/Y
∞	1995	120	−0.13
	2005	144	−0.26
1.5	1995	120	−.58
	2005	144	−.59
0.5	1995	120	0.37
	2005	144	0.32
0	1995	120	0.43
	2005	144	0.49

Y is income per worker.

values of the elasticity of substitution between reproducible and natural capital. [9]

---

[9] In the table the values corresponding to an infinite elasticity of substitution simply reproduce the benchmark case. The case of zero elasticity is approximated by setting η to 1000.

PART II

INTERPRETING TECHNOLOGY
DIFFERENCES

5

AN ENDOGENOUS TECHNOLOGY FRAMEWORK

In chapters 2, 3, and 4 we established the following patterns. Technology differences are biased toward skilled labor: holding the relative supply of skills constant, the marginal productivity of skilled labor relative to unskilled labor is higher in richer countries. Technology differences are also biased toward reproducible capital. Holding constant the relative supply of reproducible capital, the marginal productivity of reproducible capital relative to natural capital is (probably) higher in richer countries. Finally, technology differences are biased toward labor. Holding the capital-labor ratio constant, the marginal productivity of labor (measured in units of unskilled labor equivalents) relative to capital (measured in units of natural capital) is (probably) higher in richer countries. We also saw that technology differences are labor augmenting: richer countries use (unskilled equivalent) labor more efficiently than poorer countries; and perhaps capital diminishing: richer countries use (natural equivalent) capital possibly less efficiently than poorer ones.

In this chapter I present a technology-choice framework capable of rationalizing these findings. In this framework, firms in each country choose a technology characterized by a particular combination of efficiency units attached to different inputs. The optimal choice of technology depends on relative factor prices and, hence, on relative factor supplies. I first develop the analysis for a production function with only skilled and unskilled labor, in order to draw out the main intuition. I then extend the model to feature the four factors of production that I have used in the empirical framework.

To motivate the two-factor version of the model, it is useful to link back to the recent literature on skilled-biased technical change. This literature has documented substantial increases in the relative marginal productivity of skilled workers over the last few decades in the United States and in several other industrialized countries [see Autor, Katz, and Krueger (1998) for a survey]. A canonical example of skilled-biased technical change is the transition from an assembly line manned by unskilled workers, supervised by a few skilled workers, to a computer-controlled facility operated by skilled workers, where unskilled workers are at best retained as janitors (if not entirely displaced). In particular, the widely held view is that the shift from assembly line–type methods to computer-based methods is strongly skilled-labor augmenting, i.e. it leads to a big increase in the efficiency units associated with skilled workers. At the same time, since unskilled workers are demoted to janitorial roles, if not entirely displaced (to resurface elsewhere in menial jobs), it is plausible that the same shift leads to a decline in the efficiency units of unskilled workers. Declines in the efficiency units of unskilled workers over time are documented in Ruiz-Arranz (2002) and Caselli and Coleman (2002), and they are consistent with the fact that absolute wages in the lower half of the wage distribution have actually declined in the United States over much of the last few decades. I return to this literature in the third part of the book.

The switch to the computer-controlled plant is of course a choice by the firm, since it could have decided to stick to the assembly line. But the fact that rich-country producers seem largely to have embraced the switch to computer-controlled production does not mean that firms in poor countries should necessarily make the same choice. In a country that is skilled-labor abundant, such as the United States, it makes sense to expect firms to adopt more skilled-biased technologies. But in countries that are abundant in unskilled labor we may expect firms to stick to the old technology and avoid the loss in the efficient use of the abundant factor. In this case, we will observe the cross-country skill bias we document: the skilled-abundant country will have relatively high A_S/A_U compared to the unskilled-abundant country.

The model generalizes this example by simply allowing a choice from a large number of technologies, instead of just the two of the example. The basic idea is that in each country firms choose from a menu of different production methods that differ in the use they make of skilled and unskilled labor (or natural and reproducible capital, or just capital and labor). Each of these methods is a different production function.

To capture the idea that different production functions use different inputs more or less efficiently we assume that all production functions are of the form (1.7), but they differ in the parameters A_1 and A_2. Hence, we can represent the menu of possible choices of production function by a set of possible (A_1, A_2) pairs. Clearly no country will use a production function characterized by a certain pair (A_1, A_2) when another production function exists such that both A_1 and A_2 are higher, so only *nondominated* (A_1, A_2) pairs are relevant. We call this set of nondominated (A_1, A_2) pairs a "technology frontier." I illustrate a possible frontier, for the case of skilled and unskilled labor, in figure 5.1. The locus labeled A is the technology frontier for country A.

The profit-maximizing choice of production function depends of course on factor prices. Since factor prices depend on factor endowments, firms in countries with different endowments will operate different production functions. If country A is unskilled-labor abundant, skilled labor will be relatively expensive, so we might expect firms in this country to choose a technology such as the one represented by point A_a, i.e. a relatively unskilled-complementary technology. If, instead, this country is skill abundant, firms may choose a technology such as A_b. In terms of the existing literature, A_a is an *appropriate technology* for an unskilled-abundant country, while A_b is an appropriate technology for a skilled-abundant country.

Aside from the example I opened this subsection with, another way to motivate the idea of a technology frontier is suggested in

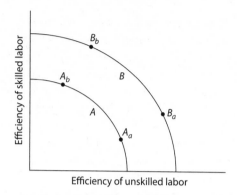

Figure 5.1. Technology Choice and Barriers to Adoption

an elegant paper by Jones (2005). Jones argues that a new invention is essentially drawn from the distribution of possible (but yet to be invented) production functions. Suppose that production functions all have the functional form (1.7) but differ in the parameters A_1 and A_2. Then a new idea—a newly invented production function—can be represented as a point in (A_1, A_2) space. Hence, technical change is nothing but the progressive "filling up" of the (A_1, A_2) space with newly available technologies. At any given point in time firms will choose their production function from this set of feasible possibilities. Clearly, again, no country will choose a dominated technology, so only the subset of nondominated production functions will be relevant. Such a set may look like a downward-sloping curve in (A_1, A_2) space: a technology frontier.

An important question is how this appropriate-technology idea can be reconciled with the (more mainstream) view that poor countries face barriers to technology adoption. This is important because, as discussed in the introductory chapter, the evidence on Total factor productivity differences is so compelling that one would not want to abandon the latter in order to embrace the former. To combine the present appropriate-technology model with the "barriers" view of technology differences, I let the technology frontiers be country specific. The idea is that countries with more severe barriers face a more limited set of choices. In figure 5.1 I illustrate this by drawing a separate frontier for country B. Since country B's frontier is higher than country A's, country B has fewer barriers to technology adoption. On its frontier, country B will choose B_a if it is unskilled-labor abundant and B_b if it is skilled-labor abundant.

The following metaphor may be helpful in thinking about the theoretical framework. Suppose that in each country there is a library, containing blueprints, or recipes to turn inputs into output. Each blueprint is associated with a different realization of the efficiency vector. For example, a blueprint entitled "computer-controlled processing" leads to high skilled-labor efficiency and low unskilled-labor efficiency and one called "assembly line" is associated with an opposite pattern of efficiencies. The different country-specific frontiers can further be interpreted as library sizes. Some countries have just a handful of blueprints that fit on a short shelf; others have roomfuls of them.

It should be clear now how combining the "appropriate technology" and the "barrier to adoption" ideas can rationalize our basic findings. Consider again the world of figure 5.1, and imagine that country A is unskilled abundant (and hence uses A_a) and country B is skilled abundant (uses B_b). If the frontiers are relatively close to each other, the appropriate-technology effect will dominate, and we will observe absolutely higher A_S in country B (the rich country) and absolutely higher A_U in country A (the poor country). This is the case depicted in the figure. If instead the frontiers are relatively far apart, the barriers effect will dominate, and A_S and A_U will both be higher in the rich country. In either case, however, the *ratio* of A_S to A_U is higher in the rich country, i.e. we always have skill bias.

I conclude this discussion by noting that the framework implicitly defines a *world technology frontier*. This can be thought of as the "highest" frontier, or the frontier of a country that faces no barriers. It represents the set of nondominated (A_U, A_S) combinations dreamed up to date by scientists and management gurus, i.e. it reflects the current state of human technical knowledge.

By introducing new technologies that dominate a subset of the preexisting ones on the frontier, technological progress shifts this locus (locally) up.[1]

The proposed model of endogenous technology choice belongs primarily in the appropriate-technology literature, which goes back at least to Atkinson and Stiglitz (1969), who called it "localized technology," and has recently been further explored theoretically by Diwan and Rodrik (1991), Basu and Weil (1998), and Acemoglu and Zilibotti (2001). The key idea in this literature that is shared by the present model is that countries with different factor endowments should choose different technologies. The Acemoglu and Zilibotti paper is particularly closely related in that it focuses on skilled and unskilled labor in order to interpret patterns in cross-country data. However, a central prediction of their model is that A_S/A_U is constant across countries, which the evidence presented in chapter 2 directly contradicts. On the empirical side, supportive evidence for appropriate technology has been developed by Caselli and Coleman (2001) and Caselli and Wilson (2004), who found that cross-country diffusion of R&D-intensive technologies is strongly influenced by factor endowments.

Like all appropriate technology models, the present one is also related to the literature on induced innovation/directed technical change, which studies the analogous problem of how factor endowments determine whether technical change will be biased toward certain factors rather than others. Important contributions in this tradition are Hicks (1932), Kennedy (1964), Samuelson (1965, 1966), Acemoglu (1998, 2002), and Jones (2005). Formally the model is closest to Samuelson's, but the argument that the cross-country skill bias documented above is driven by endogenous technology choice dictated by skilled-labor endowments

---

[1] I do not take a stand on two questions that are implicit in the foregoing discussion. First, I am agnostic about the determinants of the position of the world technology frontier in (A_U, A_S) space. Acemoglu and Zilibotti (2001) and Jones (2005) present two possible approaches to this question. Second, I am also agnostic on the sources of country-specific barriers to technology adoption.

parallels Acemoglu's (1998) idea that skilled-biased technical change in recent years is driven by endogenous responses of R&D to changes in the relative supply of skilled labor.

5.1 The Two-Factor Model

The following simple model formalizes the ideas set out in the previous subsection and establishes the conditions under which the intuition that countries will choose technologies that augment the abundant factor is valid. We will see that the key parameter is the elasticity of substitution between the two factors of production.

Consider an economy with a large number of competitive firms. Each firm generates output using a production function of the form (1.7), which I reproduce here for the special case of skilled and unskilled labor:

$$Y = [(A_U U)^\rho + (A_S S)^\rho]^{\frac{1}{\rho}} . \qquad (5.1)$$

Firms hire the two labor types taking as given the rental rates W_U and W_S. The novel element is that—besides optimally choosing factor inputs—firms also optimally choose the production function. In particular, they can choose from a menu of production functions that differ by the parameters A_U and A_S. The menu of feasible technology choices is given by

$$(A_S)^\omega + \gamma (A_U)^\omega \le B, \qquad (5.2)$$

where ω, γ, and B, all strictly positive, are exogenous parameters. This says that, on the boundary of the feasible menu—on the technology frontier—changing production function involves a trade-off between the efficiency of unskilled labor, on the one hand, and the efficiency of skilled labor, on the other. The parameters γ and ω govern the trade-off; the parameter B determines the "height" of the technology frontier. The particular functional form of equation (5.2) is dictated by technical convenience, but it is rather flexible, and it does get at the central idea that there are trade-offs associated with technology choice.

In sum, in each country the representative firm maximizes profits $(Y - W_U U - W_S S)$ with respect to U, S, and to A_U and A_S, subject to (5.1) and (5.2), the latter with equality. I close the model by assuming that the economy's endowments of U and S are all inelastically supplied. An equilibrium is a situation where all firms maximize profits and all inputs are fully employed.

In appendix A, I prove the following.

Proposition. *An equilibrium exists and is unique. If $\omega > \rho/(1 - \rho)$, the equilibrium is symmetric, in the sense that all firms choose the same technology (A_U, A_S), and the same factor ratios, S/U. If $\omega < \rho/(1 - \rho)$, the equilibrium is asymmetric, with some firms setting $A_U = 0$ and employing only skilled labor, and some others setting $A_S = 0$ and employing only unskilled labor.*

The proposition says that condition $\omega > \rho/(1 - \rho)$ is what is needed to rule out deviations from the symmetric equilibrium, deviations in which a firm chooses a corner with either $A_S = 0$ or $A_U = 0$.[2] Its meaning is rather intuitive. When ρ is low the two inputs are poor substitutes and firms will want to operate production functions with positive quantities of both S and U. But if one is going to employ both inputs, it must be the case that the respective efficiency units A_S and A_U are strictly positive. As ρ becomes larger, however, and S and U become better and better substitutes, it makes more and more sense to use only one of the inputs, and then maximize the efficiency of that input. For example, a firm may choose to set $U = 0$ and then maximize A_S by also setting $A_U = 0$. The condition says that this will happen when ρ becomes sufficiently large relative to ω. ω regulates the concavity of the technology frontier: a higher ω pushes the frontier further away from the origin, i.e. it makes interior technology choices more attractive relative to the corners. Hence, it

---

[2] Note that a symmetric equilibrium is always *interior*, in the sense that it features $A_S > 0, A_U > 0$. To see this notice that a firm choosing $A_S = 0$ $(A_U = 0)$ would also always choose $S = 0$ $(U = 0)$. But then there must be some other firm making a different technology choice.

makes firms more reluctant to move to the corners. Notice that the condition for a symmetric equilibrium is always satisfied if $\rho < 0$.

I now assume that the condition for existence of a symmetric equilibrium is satisfied, and examine this equilibrium's properties. Each firm's first-order conditions include

$$\left(\frac{S}{U}\right)^{1-\rho} = \left(\frac{A_S}{A_U}\right)^{\rho} \Big/ \frac{W_S}{W_U}, \tag{5.3}$$

$$\left(\frac{A_S}{A_U}\right)^{\omega-\rho} = \gamma \left(\frac{S}{U}\right)^{\rho}. \tag{5.4}$$

The first equation is of course just (2.1) rearranged. It combines the first-order conditions for U and S. It obviously says that the optimal choice of S/U is decreasing in W_S/W_U. For $\rho > 0$ (good substitutability between skilled labor and unskilled labor) it also says that the greater the relative efficiency of S, the greater the desired relative employment of S. For $\rho < 0$ (poor substitutability), S/U decreases in A_S/A_U, as the firm tries to boost the effective input of the inefficient (and hence effectively scarce) input.

The second equation is the first-order condition with respect to A_U. It describes how technology choice depends on the quantities of inputs employed. For $\rho > 0$, the symmetric-equilibrium condition $\omega > \rho/(1-\rho)$ implies $\omega - \rho > 0$. Hence, equation (5.4) implies that firms that employ a lot of skilled labor tend to choose technologies that augment skilled labor relative to unskilled labor. Conversely, if $\rho < 0$, firms tend to direct technology choice toward the scarce input. Now rewriting this equation as

$$\left(\frac{A_S}{A_U}\right)^{\rho} = \gamma^{\frac{1}{\omega-\rho}} \left(\frac{S}{U}\right)^{\frac{\rho^2}{\omega-\rho}}$$

we see that a country always biases its technology choices toward its relative abundant factor, in the sense that relative marginal productivities are positively correlated with relative factor supplies.

Straightforward algebra combining the first two conditions leads to the following solution to the firm's problem:

$$\frac{A_S}{A_U} = \left(\frac{W_S}{W_U}\right)^{\overline{\omega\rho-(\omega-\rho)}} \gamma^{\frac{1-\rho}{(\omega-\rho)-\omega\rho}}, \qquad (5.5)$$

$$\frac{S}{U} = \left(\frac{W_S}{W_U}\right)^{\frac{\omega-\rho}{\omega\rho-(\omega-\rho)}} \gamma^{\frac{\rho}{(\omega-\rho)-\omega\rho}}. \qquad (5.6)$$

Of course the condition $\omega > \rho/(1-\rho)$ can be rewritten as $\omega\rho - (\omega-\rho) < 0$. Hence, for $\rho > 0$ firms increase the relative efficiency of the relatively cheap factor, while for $\rho < 0$ firms focus on increasing the efficiency of the relatively expensive factor. Also, regardless of ρ, relative demand for skilled labor decreases in the relative skilled wage.

Moving from the firm's problem to the general equilibrium of the economy now is straightforward. Since the equilibrium is symmetric, equation (5.4) holds for S/U equal to the economy's endowment. Hence, with $\rho > 0$—i.e. when inputs are relatively good substitutes—countries with abundant unskilled labor will choose relatively unskilled labor–augmenting technologies, while with $\rho < 0$—or when inputs are poor substitutes—countries with abundant unskilled labor will try to boost the productivity of skilled labor. In other words, when inputs are good substitutes countries make the most of the abundant input, while when they are poor substitutes increasing the effective supply of the scarce factor is optimal. Now recall that empirically the elasticity of substitution $1/(1-\rho)$ is greater than 1, implying that $\rho > 0$. Equation (5.4)—together with the fact that U/S is higher in poor countries—is therefore the rationalization of our basic finding or skill bias.

Indeed, if all countries shared the same technology frontier, i.e. if B was the same in all countries, it would follow directly from (5.4) and (5.2) that A_U should always be absolutely higher in poor countries. However, the central message of the barriers-to-adoption literature is surely right: there are impediments to the diffusion of technology across countries. As already mentioned one can nest this idea in the model by allowing the technology frontier in equation (5.2) to be country-specific. In particular, suppose that the height of the frontier, B, varies from country to country. It is straightforward to show that in this case one gets skill bias—it's equation (5.4)!—without necessarily implying that absolute unskilled efficiency is higher in poor countries. In particular, if B is much higher in rich countries, the absolute levels of both A_S and A_U will be higher in those countries. This can be seen formally by combining equations (5.4) and (5.2) to get

$$A_S = \left(\frac{B}{1 + \gamma^{\rho/(\rho-\omega)}(S/U)^{\omega\rho/(\rho-\omega)}} \right)^{1/\omega}, \qquad (5.7)$$

$$A_U = \left(\frac{B/\gamma}{1 + \gamma^{\rho/(\omega-\rho)}(S/U)^{\omega\rho/(\omega-\rho)}} \right)^{1/\omega}. \qquad (5.8)$$

Recalling that $\omega > \rho$ is implied by our condition for an interior optimum, this says that A_S is increasing in both B and S/U, while A_U is increasing in B and decreasing in S/U (as long as $\rho > 0$).

5.2 The Four-Factor Model and the Evidence

It is straightforward to generalize the two-factor model to feature the four factors I used in the empirical analysis. The problem faced by each country's firms is now

$$\max_{U,S,M,N,A_S,A_U,A_M,A_N} \left\{ [(A_U U)^\rho + (A_S S)^\rho]^{\sigma/\rho} \right.$$

$$+ [(A_M M)^\eta + (A_N N)^\eta]^{\sigma/\eta} \Big\}^{1/\sigma}$$

$$- W_u U - W_S S - R_M M - R_N N$$

subject to: $\gamma_S (A_S)^\omega + \gamma_U (A_U)^\omega + \gamma_M (A_M)^\omega + \gamma_N (A_N)^\omega \leq B$

where the production function is the production function I used in the first part of the book, and the technology frontier has the same interpretation as in the two-factor model, but now features a choice among four augmentation coefficients.

Combining the first-order conditions with respect to A_S and A_U we obtain

$$\left(\frac{A_S}{A_U}\right)^{\rho} = \left(\frac{\gamma_U}{\gamma_S}\right)^{\frac{\rho}{\omega-\rho}} \left(\frac{S}{U}\right)^{\frac{\rho^2}{\omega-\rho}}, \qquad (5.9)$$

which is the identical result to the two-factor model. Our finding that $(A_S/A_U)^{\rho}$ is increasing in income per worker can be rationalized if richer countries have a greater relative supply of skills, S/U. Of course we already know this is true from chapter 2 (see figure 2.1).

By the same token, we get

$$\left(\frac{A_M}{A_N}\right)^{\eta} = \left(\frac{\gamma_U}{\gamma_S}\right)^{\frac{\eta}{\omega-\eta}} \left(\frac{M}{N}\right)^{\frac{\eta^2}{\omega-\eta}}. \qquad (5.10)$$

Our (tentative) conclusion that $(A_M/A_N)^{\eta}$ is increasing in income per worker can be rationalized if richer countries have a greater relative supply of skills, M/N. Of course we already know this is true from chapter 3 (see figure 3.1).

In chapter 4 we also made inferences about the cross-country behavior of $\tilde{A}_L = A_L A_U$ and $\tilde{A}_K = A_K A_N$. As discussed in section 1.6 we cannot separately identify A_L from A_U and A_K from A_N. Accordingly I have normalized A_L and A_K to 1 in this section. Hence, $\tilde{A}_L = A_U$ and $\tilde{A}_K = A_N$. Using again the first-order conditions with respect to A_N and A_U, we have

$$\tilde{A}_K = A_N = \left(\frac{Y^{1-\sigma} K^{1-\eta} N^{\eta}}{\lambda \gamma_N \omega}\right)^{\frac{1}{\omega-\eta}}, \qquad (5.11)$$

$$\tilde{A}_L = A_U = \left(\frac{Y^{1-\sigma} L^{1-\rho} U^{\rho}}{\lambda \gamma_U \omega}\right)^{\frac{1}{\omega-\rho}}, \qquad (5.12)$$

and

$$\left(\frac{\tilde{A}_K}{\tilde{A}_L}\right)^\sigma = \left(\frac{A_N}{A_U}\right)^\sigma$$

$$= \left[\frac{(\gamma_U)^{\frac{1}{\omega-\rho}}}{(\gamma_N)^{\frac{1}{\omega-\eta}}}\right]^\sigma \left[\frac{(K^{1-\eta}N^\eta)^{\frac{1}{\omega-\eta}}}{(L^{1-\rho}U^\rho)^{\frac{1}{\omega-\rho}}}\right]^\sigma \left(\frac{Y^{1-\sigma}}{\lambda\omega}\right)^{\frac{\sigma(\eta-\rho)}{(\omega-\eta)(\omega-\rho)}}.$$

$$(5.13)$$

The predictions in (5.11)–(5.13) are harder to assess empirically, because they depend on the sign and magnitude of the elasticities σ, η, and ρ. As discussed in part I of the book, there is little consensus on the first two. They also depend on the value of the unknown parameter ω.

I fall back on a more heuristic approach. As discussed, the intuition behind (5.9) and (5.10) is that technology choice is biased toward the more abundant factor. Within this logic, our (tentative)

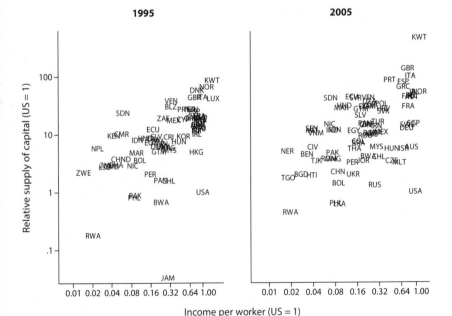

Figure 5.2. $\tilde{K}/\tilde{L}$ against $\tilde{Y}$

conclusion that $(\tilde{A}_K/\tilde{A}_L)^\sigma$ is increasing in income per worker might be rationalized if richer countries have a greater relative (quality-adjusted) supply of capital, $\tilde{K}/\tilde{L}$. As discussed in chapter 4 the measurement of $\tilde{K}$ is fraught with uncertainty. With that caveat, figure 5.2 shows that $\tilde{K}/\tilde{L}$ is indeed positively associated with income per worker when I use my benchmark values for both $\tilde{L}$ and $\tilde{K}$.

Another prediction of the two-factor model is that (as long as the "intercept" B does not vary too much across countries) we should observe $\tilde{A}_K$ ($\tilde{A}_L$) increase (decrease) in $\tilde{K}/\tilde{L}$ if $\sigma > 0$ and decrease if $\sigma < 0$. Empirically, we found that $\tilde{A}_K$ may be decreasing in income, and hence in $\tilde{K}/\tilde{L}$, and that $\tilde{A}_L$ is increasing in income and $\tilde{K}/\tilde{L}$. Since we also argued that $\sigma < 0$ is likely to be the empirically relevant case, these findings are also rationalized by the model.

PART III

Technology Differences over Time

6

SKILLED LABOR, UNSKILLED LABOR, AND EXPERIENCE OVER TIME

6.1 Introduction

The relative supplies, and relative rewards, of workers with different characteristics are constantly changing. As in the case of cross-country comparisons, changes in rewards are explained partially by changes in relative supplies and partially by nonneutral changes in technology. The purpose of this chapter is to apply the same techniques that were used in part I to investigate what the joint behavior of relative wages and relative supplies reveal about the underlying changes in technology. For ease of access to data and comparability to the existing literature I focus the analysis on the United States.

I will distinguish workers by two characteristics: skill and experience. In other words, I allow for a fourfold partition of the labor force: experienced skilled workers, inexperienced skilled workers, experienced unskilled workers, and inexperienced unskilled workers.

As noted in chapters 1 and 2, the skilled-unskilled dichotomy is the object of a large literature in labor economics and macroeconomics. While the literature on this topic has always been active, interest peaked during the 1990s, in response to a spectacular increase in the college wage premium—the ratio of wages received by "college-graduate equivalents" relative to "high school–graduate equivalents." Most of the authors who have investigated these changes agree that nonneutral changes in technology biased toward college graduates—known as skilled-biased technical change (SBTC)—are an important part of the story. Here I will revisit and update this conclusion using the techniques already deployed in the cross-country context.

The distinction between experienced and inexperienced workers and, in particular, the possibility of an "experience" bias in technical change are less common in the literature. In their classic 1992 paper Katz and Murphy discussed this possibility, but their analytical framework, unlike mine, was not able to distinguish between SBTC and experience-biased technical change. I am motivated here to revisit this topic by evidence in Card and Lemieux (2001), Guvenen and Kuruscu (2010), Acemoglu and Autor (2011), and others, that show marked differences in the behavior of college premia for younger and older workers. The techniques of the book may perhaps help shed light on these differences.[1]

I will work with the following generalization of the functional form for the composite labor input:

$$\tilde{L}_t = \left\{ \left[U I_t^{\eta_U} + (A_{UEt} U E_t)^{\eta_U} \right]^{\rho/\eta_U} \right.$$

$$\left. + A_{St}^{\rho} \left[S I_t^{\eta_S} + (A_{SEt} S E_t)^{\eta_S} \right]^{\rho/\eta_S} \right\}^{1/\rho} . \quad (6.1)$$

In this representation, which is essentially Card and Lemieux's, UI, UE, SI, and SE denote the quantities of unskilled inexperienced inputs, unskilled experienced inputs, skilled inexperienced

---

[1] Recently, Jeong et al. (2015) conclude that there is no need for demand shifts to explain changes over time in the "price of experience" in the United States. However, their conceptual framework is very different from mine and their notion of the price of experience does not match well with the experience premium analyzed here. They postulate an aggregate production function defined on two inputs: a "pure labor" input and an "experience" input. The price of experience is the relative price between these two. In my framework the production function is defined over four inputs: experienced/inexperienced high school/college graduates. The experience premia are the relative wages of experienced workers. It is perhaps possible to argue that my framework, being defined in terms of bodies, poses fewer measurement challenges than the one founded on the abstract notions of the overall supply of "pure labor" and "experience." Interpretation is also perhaps a bit more straightforward. Boehm and Siegel (2014) combine Jeong et al.'s framework with a panel-IV strategy. Unlike Jeong et al. their preliminary results do show a significant role for demand shifts.

inputs, and skilled experienced inputs, respectively. The time-invariant coefficients η_U and η_S govern the elasticity of substitution between unskilled inexperienced and unskilled experienced workers, and skilled inexperienced and skilled experienced ones, respectively. The parameter ρ continues to govern the elasticity of substitution between unskilled and skilled workers. Finally, the time-varying coefficients A identify nonneutralities in technological change: $A_{UEt}^{\eta_U}$, and $A_{SEt}^{\eta_S}$ capture the "experience bias" within the unskilled and the skilled group, respectively; A_{St}^{ρ} captures the skill bias, and has an interpretation identical to $(A_{Sc}/A_{Uc})^{\rho}$ in the cross-country context. The goal is to characterize the time series behavior of these As.

6.1.1 Data

As explained in more detail below, backing out the As that appear in equation (6.1) requires time series data on the labor supplies UI_t, UE_t, SE_t, SI_t, and the corresponding wages w_{UIt}, w_{UEt}, w_{SEt}, w_{SIt}. I construct these series from data developed by Acemoglu and Autor (2011), henceforth AA, using the 1963–2008 March CPS samples.

AA make available a variable measuring total annual hours of labor by gender, 5 education categories, and 48 experience categories (i.e. from 0 to 48 years of experience). I define as "inexperienced" all workers with 19 years of experience or less, and "experienced" those with 20–48 years of experience.[2] I further define as "unskilled" all high school dropouts, high school graduates, and workers with incomplete college (education categories 1–3). The "skilled" are those with a college or a postgraduate degree (education categories 4–5).

AA also compute the average weekly full-time equivalent earnings within each of these gender-education-experience cells. I pick male high school graduates with 10 years of experience as the

---

[2] Nineteen years of experience is the (hours of labor supply–weighted) average over time in the CPS (the unweighted average is 22). Corresponding medians are 18 and 22.

reference group for the unskilled-inexperienced category, male high school graduates with 30 years of experience as benchmark for the unskilled-experienced group, and male college graduates with 10 and 30 years of experience as baseline for skilled-inexperienced and skilled-experienced, respectively. Then, for each gender-education-years of experience cell I construct a fixed weight given by mean earnings in that cell relative to the relevant benchmark mean earnings, averaged over the sample period. The idea of these weights is that they represent an efficiency-unit conversion factor to express hours supplied by a given cell into hours supplied by the reference cell within the education-experience category. Using these weights, I construct UI_t, UE_t, SE_t, SI_t as weighted sums of the hours supplied by each gender-education-years of experience cell within each of the four broad education-experience categories.

Figure 6.1 plots the time series of the labor supplies, in logs and normalized by their value at the beginning of the sample period. The behavior of the various labor supply series is dominated by the rise in skills, with larger fractions of each cohort achieving college degrees. But also important are demographics, and particularly the baby-boom cycle. Hence, up to the end of the 1980s we tend to see faster growth in the inexperienced groups, and thereafter in the experienced ones as the baby-boom generation transitions from one to the other. As a result, the grouping that has experienced the largest increase over the sample period is SE, followed by SI.

For each annual data set, AA also regress individual log weekly wages on five dummies corresponding to the five levels of educational attainment, a quartic in experience, a gender dummy, a race dummy, and several interactions of these variables. They then construct predicted real log weekly wage series for white workers by gender, five educational attainment categories, and five levels of experience, namely, 5, 15, 25, 35, and 45 years. I simply take the predicted wage series for high school graduates (college graduates) as representative of the unskilled (skilled) category, and the series for workers with 5 (25) years of experience as representative

Figure 6.1. Labor Supply by Skill and Experience

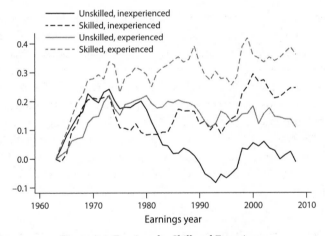

Figure 6.2. Earnings by Skill and Experience

of the inexperienced (experienced) category. This gives me (logs of) w_{UIt}, w_{UEt}, w_{SEt}, w_{SIt}. These are plotted in figure 6.2 (normalized by their initial value). The pattern that dominates figure 6.2 is the well-known stagnation of real labor incomes, particularly (but not exclusively) for the unskilled.

6.2 Methodology

6.2.1 Estimating Elasticities of Substitutions between Experienced and Inexperienced Workers

I begin by obtaining estimates of the elasticity of substitutions between experienced and inexperienced workers. Assuming perfectly competitive labor markets, we can derive the following formulas for the *experience premia* :

$$\frac{w_{UEt}}{w_{UIt}} = A_{UEt} \left(\frac{UE_t}{UI_t} \right)^{\eta_U - 1}, \tag{6.2}$$

$$\frac{w_{SEt}}{w_{SIt}} = A_{SEt} \left(\frac{SE_t}{SI_t} \right)^{\eta_S - 1}. \tag{6.3}$$

Figure 6.3 plots experience premia and relative supplies of experience for unskilled (left panel) and skilled workers (right panel). In both cases, relative supplies follow a deep U-shaped pattern driven by the baby boom (as discussed above). However, experience premia do not appear hugely responsive, particularly for skilled workers. This suggests a fairly large elasticity of substitution between experienced and inexperienced workers.

Nevertheless, the two elasticities of substitution are difficult to identify, as the experience biases A_{UE} and A_{SE} are unobservable. However, the elasticities of substitution can be identified if we assume

$$A_{UEt} = \chi A_{SEt} + \omega_t,$$

where ω_t is i.i.d. In other words, we assume that the experience bias has a common trend for skilled and unskilled workers—presumably a fairly plausible assumption. With this assumption, we can combine the two expressions for the experience premium into a Diff-inDiff specification:

$$\log \frac{w_{SEt}}{w_{SIt}} - \log \frac{w_{UEt}}{w_{UIt}} = \alpha + (\eta_S - 1) \log \frac{SE_t}{SI_t} - (\eta_U - 1) \log \frac{UE_t}{UI_t} + \varepsilon_t,$$

which can be estimated by OLS.

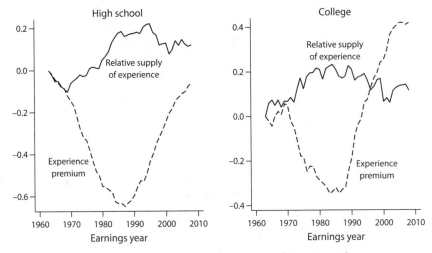

Figure 6.3. Relative Prices and Quantities of Experience, by
Educational Attainment

The OLS coefficients (standard errors) from this regression are
−0.342 (0.046) and 0.303 (0.054). These imply that the elasticities
of substitutions are

$$\frac{1}{1 - \eta_U} = 3.3 \quad \text{and} \quad \frac{1}{1 - \eta_S} = 2.9,$$

with standard errors 0.586 and 0.392, respectively.[3] Not surprisingly, the experienced-inexperienced elasticities of substitution
are quite high.

6.2.2 Backing Out Experience Biases

With estimates of the elasticities η_S and η_U at hand, we can
return to equations (6.2) and (6.3) and solve them for the experi-
ence biases $A_{UEt}^{\eta_U}$ and $A_{SEt}^{\eta_S}$. These are plotted (in logs) in figure
6.4. The figure reveals marked positive trends in the experience

---

[3] The fit of the regression is reasonable, with an R-squared of 0.57 and a mean
square error of 0.05.

Figure 6.4. Experience Bias by Skill

bias for both unskilled and skilled workers, particularly since 1980. Such experience-biased technical change is likely to have occurred within industries and occupations, as automation (and other advances) have diminished the requirements for physical strength and stamina (and possibly increased the benefits of experience). At the aggregate level, experience-biased technical change may also be the reduced-form implication of structural changes that have diminished the weight of sectors where workers perform physical tasks, such as manufacturing.

To understand this result, refer back to figure 6.3. Since 1980 or so, the relative supply of experience has increased very markedly (in both skill groups), and yet experience premia have hardly declined. Even with our relatively large estimated elasticities of substitution between skilled and unskilled workers, the experience premium relative stability in the face of a large increase in the relative supply of experience must imply that technological change has been experience biased. Before 1980, the relative supply of skills was declining in both groups, and both groups duly experienced an increase in the experience premium. For the unskilled, the increase in the premium was roughly what one would expect given the estimate of η_U, so the experience bias is relatively flat. For the skilled, the increase in the premium is actually *greater* than what one would expect given η_S, leading to a positive trend in the skilled experience bias for the early sub-period as well.

6.2.3 Backing Out the Skill Bias

The skill (or college) premia for inexperienced and experienced workers are given by

$$\frac{w_{SIt}}{w_{UIt}} = A_{St} \frac{\left[SI_t^{\eta_S} + A_{SEt} SE_t^{\eta_S} \right]^{\frac{\rho}{\eta_S} - 1} SI_t^{\eta_S - 1}}{\left[UI_t^{\eta_U} + A_{UEt} UE_t^{\eta_U} \right]^{\frac{\rho}{\eta_U} - 1} UI_t^{\eta_U - 1}}, \qquad (6.4)$$

$$\frac{w_{SEt}}{w_{UEt}} = A_{St} \frac{\left[SI_t^{\eta_S} + A_{SEt} SE_t^{\eta_S} \right]^{\frac{\rho}{\eta_S} - 1} A_{SEt} SE_t^{\eta_S - 1}}{\left[UI_t^{\eta_U} + A_{UEt} UE_t^{\eta_U} \right]^{\frac{\rho}{\eta_U} - 1} A_{UEt} UE_t^{\eta_U - 1}}. \qquad (6.5)$$

Hence, we can back out the skill bias A_S either from data on college premia among experienced workers, coupled with data on relative supplies *augmented* with our estimates of the (skilled) experience bias, or from data on college premia among inexperienced workers. The only additional input required is an estimate of the elasticity of substitution between skilled and unskilled workers, $1/(1 - \sigma)$. As elsewhere in the book, we rely on microeconomic estimates of this elasticity that put it at 1.5.

As noted by Card and Lemieux (2001), the skill premium for a given experience group depends on (i) the overall skill bias in technology, (ii) the overall relative supply of skills, and (iii) the relative supply of skills specific to the given experience group. What I have added here is that technology can have an experience bias, whereas Card and Lemieux only allow for a skill bias.

In figure 6.5 I plot the experience-specific skill premium, the overall relative supply of skills, which I define as

$$\frac{\left[SI_t^{\eta_S} + A_{SEt} SE_t^{\eta_S} \right]^{1/\eta_S}}{\left[UI_t^{\eta_U} + A_{UEt} UE_t^{\eta_U} \right]^{1/\eta_U}},$$

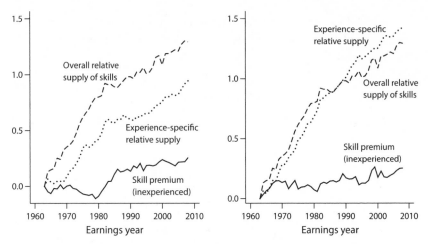

Figure 6.5. Experience Premium by Skill and Its Components

and the experience-specific relative supply of skills SI/UI and SE/UE for inexperienced (left panel) and experienced (right panel) workers. As Card and Lemieux point out, the skill premium for both experience groups has increased overall during the sample period, but the timing of the increases are quite different, with the skill premium among the inexperienced rising rapidly since the 1980s while that for the experienced taking off later and more gradually. The figures also lend credence to Card and Lemieux's conclusion that these different patterns can be understood by noting that the relative experience-specific relative supply of skills has grown faster and more sharply for the experienced.

But the key trend that dominates both panels is obviously the fact that both skill premia have risen in the context of a simultaneous large increase in the relative supply of skills, both overall and experience-group specific. It is this observation that spawned the SBTC literature and led to the conclusion that there must be a positive trend in A_s^ρ. This is confirmed in figure 6.6, which plots the two variants of A_s backed out from equations (6.4) and (6.5), respectively. The skill bias implied by the experienced skill premium shows a larger increase than the skill bias implied by the

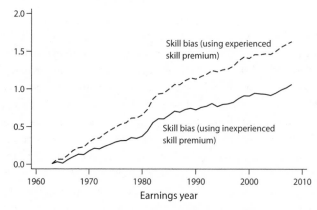

Figure 6.6. Skill-Biased Technical Change

inexperienced skill premium (which clearly implies that the model does not fit the data quite perfectly), but clearly both cases show a pronounced upward trend.

6.2.4 Summing Up

In sum, I confirm many previous findings of a significant skill bias in technical change over the last 50 years. In addition, I present (novel, I believe) evidence of an experience bias in technical change over roughly the same period, especially among skilled workers and since the 1980s.

The theoretical framework from part II, which we used to interpret the cross-country findings, can similarly be applied to the time series results. Given the very large increase in the relative supply of skills, coupled with the observation that the elasticity of substitution between skilled and unskilled workers is greater than 1, it is not surprising to observe that firms have adopted increasingly skill-biased technologies. Of course the theoretical framework takes the menu of technologies as given, which is unsatisfactory in the context of historical changes. Acemoglu (1998, 2002) shows how to endogenize the skill bias in a context of R&D-based technical change.

The data also show a substantial increase in the relative supply of experience since the early 1980s, coinciding with the maturing of the baby-boom generation. Since we estimate the elasticity of substitution among experienced and inexperienced workers to be high, it is again consistent with the theoretical framework that there would be a pronounced experience bias in this subperiod. However, the supply of experience declined in the 1960s and 1970s, and we do not observe a corresponding decline in the relative efficiency of experienced labor over this earlier period. One possible explanation, though, is that the subsequent, demographic-driven, reversal in relative supplies was predictable. Firms that were aware of the coming acceleration in the relative supply of experience would probably not have wanted to temporarily switch to inexperienced-biased technologies.

7

SKILLS AND CAPITAL OVER TIME AND ACROSS COUNTRIES

7.1 Introduction

Up to now I have applied my techniques to identifying factor biases in technology either to a broad cross section of countries or to a time series for a single country. In this chapter I bring the approach to a small panel of (industrialized) countries for which minimal data requirements are met. The intent is twofold: first, to investigate whether the trends in skill bias observed in the United States are common to other economies; second, to extend the time series analysis to include capital, whereas until now it has been limited to technology biases among types of workers.

The conceptual framework is given by equations (1.4) and (1.6), which I reproduce here for convenience:

$$Y_{ct} = [(A_{Kct}K_{ct})^\sigma + (A_{Lct}L_{ct})^\sigma]^{1/\sigma}, \qquad (7.1)$$

$$L_{ct} = [(A_{Uct}U_{ct})^\rho + (A_{Sct}S_{ct})^\rho]^{1/\rho}.$$

There is no distinction between natural and reproducible capital, but natural capital represents a relatively small share of the capital input in the countries in the sample with which I work in this chapter. I also do not distinguish workers by experience, mostly in order to keep the framework relatively simple.

The exercise is by now familiar. I will first back out, for each country, the skill bias $(A_{Sct}/A_{Uct})^\rho$ from data on the relative supply of skills and the relative wages of skilled workers. With the estimated skill bias at hand, I construct labor supply in units of equivalent unskilled workers as

$$\tilde{L}_{ct} = \left[(U_{ct})^{\rho} + \left(\frac{A_{Sct}}{A_{Uct}} S_{ct} \right)^{\rho} \right]^{1/\rho},$$

which I then plug into equation (7.1). Finally, I use data on overall labor and capital shares to back out the augmentation coefficients A_{Kct} and $\tilde{A}_{Lct} = A_{Lct} A_{Uct}$ from equations (4.2) and (4.3), which I reproduce:

$$\tilde{A}_{Lc} = \left(\frac{\tilde{W}_c \tilde{L}_c}{Y_c} \right)^{1/\sigma} \frac{Y_c}{\tilde{L}_c},$$

$$\tilde{A}_{Kc} = \left(\frac{\tilde{R}_c \tilde{K}_c}{Y_c} \right)^{1/\sigma} \frac{Y_c}{\tilde{K}_c}.$$

7.2 Data

The source for the panel-data approach is EU-KLEMS [O'Mahony and Timmer (2009)], which reports time series for output, capital, different types of labor, compensation to factors, and several other variables for 13 industrialized countries.

For Y I use the KLEMS series for *gross value added*. For K I use *capital services*. Regarding labor types, KLEMS breaks them up into three categories: low, medium, and high skill. Because of differences in education systems, the boundaries between these categories are not perfectly comparable across countries. The comparability problem is especially severe between low and medium skill, while the definition of high skill maps fairly consistently into having a university degree or higher. For this reason, and also for consistency with the time series analysis for the United States, I lump the low- and medium-skill categories into U and reserve the notation S for the high-skill KLEMS measure of labor supply. In aggregating low- and medium-skill workers to form the U aggregate I weigh medium-skill workers by their (country-specific mean) wage relative to low-skill workers, as in the previous chapter. KLEMS labor supplies are measured in hours.

I am able to generate time series estimates of A_S/A_U and $\tilde{A}_L$ for 24 countries for a maximum of 36 years and a minimum of 11 years. I can generate estimates of A_K for 22 countries, again with time series observations varying from 11 to 36.[1]

7.3 Results

The results for $\log(A_S/A_U)^\rho$ are depicted in figure 7.1. Skill-biased technical change emerges as a remarkably global phenomenon. Not a single country fails to register a positive trend in the relative efficiency of skilled labor.[2]

To produce estimates for A_K and $\tilde{A}_L$ I need an estimate of the elasticity of substitution $1/(1-\sigma)$. As already noted at several points, there is considerable uncertainty about this parameter, but most authors lean toward the conclusion that it is less than 1. I use 0.5 as my benchmark estimate. The time series paths of $\log(\tilde{A}_S)$ and $\log(A_K)$ implied by this choice of σ are shown in figures 7.2 and 7.3, respectively. Each country's time series is normalized to its end-of-sample-period value.[3] In virtually all countries there is a positive trend in the efficiency of the labor aggregate and a *negative* trend in the efficiency of capital. The latter result would seem very surprising had we not already encountered its analog in the cross section, where richer countries appear to use capital less efficiently. It does appear that during the growth process countries trade the efficiency with which they use labor with the efficiency with which they use capital.

---

[1] When necessary and appropriate I splice the data for West Germany and postunification Germany.

[2] At the same time, the *cross-sectional* relation between relative efficiency of skilled labor and relative supply of skilled labor is the same that we found in the broader cross section: in all years skill-abundant countries use skilled labor more efficiently.

[3] I normalize the data because the levels of A_K and $\tilde{A}_L$ are not comparable across countries. I use the end-of-sample value because all countries' sample periods end in the same year, while the beginning date varies wildly.

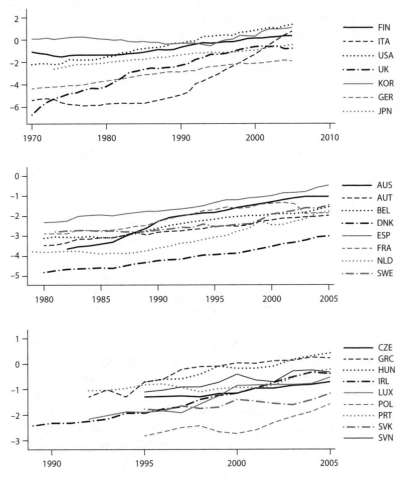

Figure 7.1. Time Series of log $(A_S/A_U)^\rho$ for OECD Countries

To check that these results are consistent with the theoretical framework in part II, table 7.1 presents summary statistics about the correlation between the labor bias in technical change, $\tilde{A}_S/A_K$, and the relative supply of labor $\tilde{L}/K$. In line with the model's predictions, in all cases bar one there is a strong

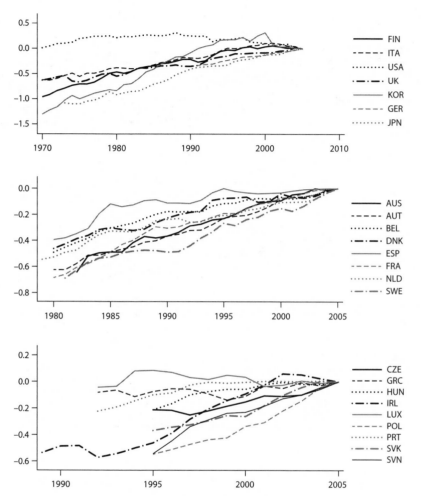

Figure 7.2. Time Series of $\log(\tilde{A}_L)$ for OECD Countries

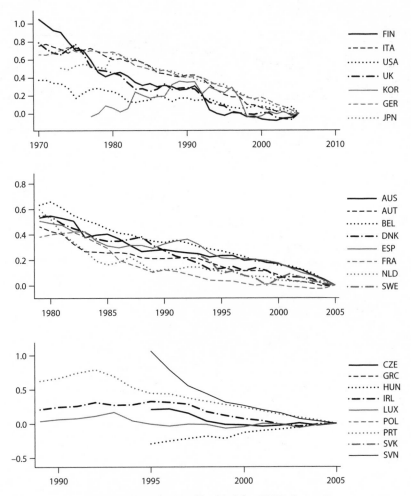

Figure 7.3. Time Series of $\log(A_K)$ for OECD Countries

TABLE 7.1. Correlations between $\tilde{A}_{Lc}/A_{Kc}$
and $\tilde{L}_c/K_c$ for OECD Countries

Country	Corr($\tilde{A}_L/A_K, \tilde{L}_c/K_c$)	Obs
AUS	−0.94	24
AUT	−0.95	26
BEL	−0.95	26
CZE	−0.87	11
DEW	−0.93	22
DNK	−0.94	26
ESP	−0.92	26
FIN	−0.95	36
FRA	−0.93	26
GER	−0.92	15
HUN	0.46	11
IRL	−0.92	18
ITA	−0.93	36
JPN	−0.95	33
KOR	−0.90	29
LUX	−0.37	14
NLD	−0.92	27
PRT	−0.90	14
SVN	−0.89	11
SWE	−0.87	13
UK	−0.96	36
USA	−0.85	36

negative correlation.[4] When the elasticity of substitution between two inputs is less than 1, technology choice shifts toward the input that becomes more scarce. In the OECD, K has been growing faster than $\tilde{L}$, so the bias in technology has favored labor.[5]

---

[4] The one exception, Hungary, has only 11 observations.

[5] Using an elasticity of substitution greater than 1 (e.g. 1.5) leads to a much less systematic pattern: in 12 countries the correlation between relative supplies and relative effciency is negative (which would be inconsistent wiht the theory if the elasticity was indeed greater than 1), while in 8 it is positive.

8

CONCLUSIONS

The aggregate production function is a central tool of most work in macroeconomics. Most of this work is predicated on a rather inflexible view of how the production function changes across countries and over time. This view is "linear" : if a country is $x\%$ more efficient at using one factor of production than another, then it uses all factors $x\%$ more efficiently. If a country experiences an $x\%$ improvement in the efficiency with which it uses one factor, then all of its factors' efficiencies improve by $x\%$

The results presented in these pages imply that technology and technical change are more flexible than usually allowed. The efficiency of different factors changes across countries and over time at different rates. Indeed, in some instances the efficiency with which one factor is used can decline while the efficiency of others increases.

Since the 1990s, it has been increasingly clear that technical change tends to have a skill bias. But the evidence from this book shows that nonneutralities are much more pervasive than that. They also occur across countries, and not just over time. And they invest a broader set of inputs: not only skilled and unskilled labor, but also experienced and inexperienced workers, natural and reproducible capital, and a broad labor aggregate and a broad capital aggregate.

The existence of marked nonneutralities in technology, and even trade-offs in the efficiencies with which different factors are used, should not be surprising. Different countries have different factor endowments, and factor endowments change over time in a given country. In a venerable tradition of economic models, countries endogenously adapt their technology to their factor endowments, and R&D efforts are redirected in response to changes in the supply of different factors. Once again, the mounting evidence of skilled-biased technical change since the 1990s has rekindled some attention to this class of models. Here I have shown how models along these lines can also be useful in understanding patterns of nonneutrality among different types of experience groups, different types of capital, and between labor and capital. The models can also potentially explain why the efficiency with which some factors are used must sometimes fall to let the efficiency of other factors increase.

Needless to say, this book merely scratches the surface of the likely patterns of nonneutrality that exist across countries and over time. To begin with, some of the conclusions presented here are tentative, as they are predicated on parameter values for which our knowledge is approximate at best, particularly the aggregate elasticity of substitution between reproducible and natural capital, and (perhaps to a lesser extent, as long as we accept the widely held view that it is less than 1) the aggregate elasticity of substitution between labor and capital.

More fundamentally, despite that fact that it relies on more disaggregated measures of the factors of production than is customary, the present analysis still relies on very strong assumptions on the substitutability of types within these measures. For example, high school graduates are assumed to be perfect substitutes with primary school graduates; equipment is a perfect substitute with structures. As soon as our knowledge of patterns of substitutability within these aggregates improves, it will be possible to uncover an even richer web of nonneutralities in technology differences.

Finally, the analysis in this book has been entirely focused at the country level as the basic unit of analysis. Immense progress on understanding nonneutrality in technology could come from industry- and (even better) firm-level applications of the techniques described in this book.

APPENDIX A: PROOFS AND CALCULATIONS

Begin with equation (2.8):

$$
b = \frac{\sum_{j\leq 4}(\log W_U + \beta_j)(s_j - \mu_s)l_j + \sum_{j>4}(\log W_S + \beta_j)(s_j - \mu_s)l_j + \sum_i \varepsilon_i(s_i - \mu_s)}{\sum_j (s_j - \mu_s)^2 l_j}
$$

$$
= \frac{\log W_U \sum_{j\leq 4}(s_j - \mu_s)l_j + \sum_{j\leq 4}\beta_j(s_j - \mu_s)l_j + \log W_S \sum_{j>4}(s_j - \mu_s)l_s + \sum_{j>4}\beta_j(s_j - \mu_s)lj}{\sum_j (s_j - \mu_s)^2 l_j}
$$

$$
= \frac{\log W_U \sum_{j\leq 4}(s_j - \mu_s)l_j + \log W_S \sum_{j>4}(s_j - \mu_s)l_s + \sum_j \beta_j(s_j - \mu_s)l_j}{\sum_j (s_j - \mu_s)^2 l_j}
$$

$$
= \frac{\log W_U \sum_{j\leq 4}(s_j - \mu_s)l_j + \log W_U \sum_{j>4}(s_j - \mu_s)l_j + \log W_S \sum_{j>4}(s_j - \mu_s)l_s - \log W_U \sum_{j>4}(s_j - \mu_s)l_j + \sum_j \beta_j(s_j - \mu_s)l_j}{\sum_j (s_j - \mu_s)^2 l_j}
$$

$$
= \frac{\log W_U \sum_{j}(s_j - \mu_s)l_j + (\log W_S - \log W_U) \sum_{j>4}(s_j - \mu_s)l_j + \sum_j \beta_j(s_j - \mu_s)l_j}{\sum_j (s_j - \mu_s)^2 l_j}
$$

$$= \frac{(\log W_S - \log W_U) \sum_{j>4}(s_j - \mu_s)l_j + \sum_j \beta_j(s_j - \mu_s)l_j}{\sum_j (s_j - \mu_s)^2 l_j}.$$

The last expression solves for equation (2.9).

A.2 Existence and Uniqueness of Symmetric Equilibrium

Consider first the optimal choice of inputs for a firm that faces given factor prices W_S and W_U, and has a given technology A_U, A_S. The solution to the cost-minimization problem can be shown to give rise to the following cost function:

$$\text{Cost}(W_U, W_S; Y) = \left[\left(\frac{W_U}{A_U}\right)^{\frac{-\rho}{\rho-1}} + \left(\frac{W_S}{A_S}\right)^{\frac{-\rho}{\rho-1}} \right]^{\frac{\rho-1}{\rho}} Y.$$

Note that this cost function also accurately describes minimized costs when A_U or A_S is zero. Now it is obvious that even if A_U and A_S are chosen by the firm, the choice of factors must still be cost minimizing in the above sense. Furthermore, since the cost function is linear in output the optimal choice of technology must itself be cost minimizing. Hence, the choice of an optimal technology is a choice of (A_U, A_S) on a country's technology frontier that minimizes this cost function.

Make the change of variables $D_u = (A_U)^\omega$ and $D_s = (A_S)^\omega$. To simplify the notation, also write $\theta = \rho/\omega(1-\rho)$. We can then write the firm's problem as

$$\text{Min}_{\{D_s, D_u\}} \Big\{ \text{Cost}(W_U, W_S; Y)$$

$$= \left[(W_U)^{\frac{\rho}{\rho-1}} (D_u)^\theta + (W_S)^{\frac{\rho}{\rho-1}} (D_s)^\theta \right]^{\frac{\rho-1}{\rho}} Y \Big\},$$

$$\text{Subject to}: D_s + \gamma D_u = B.$$

Consider first the case where $\theta < 1$, or $\omega > \rho/(1-\rho)$. It is clear in this case that the firm's problem has a unique interior solution. Hence, if this condition is satisfied, all firms choose the same interior technology. The particular technology choice depends on factor prices. From the first-order conditions for an interior optimum, we have (5.6)—which shows that if firms are in a symmetric equilibrium, there is a unique equilibrium wage ratio for given S/U. Hence, we have existence and uniqueness in the $\theta < 1$ case.

For the $\theta > 1$ case it is immediate that the firm cost-minimization problem requires firms to be at a corner, with either $A_S = S = 0$ or $A_U = U = 0$. The zero-profit condition for firms choosing the former strategy is $\left(W_U/(B/\gamma)^{1/\omega}\right) = 1$, where the left-side term is the unit production cost and the right is the unit revenue. Similarly, for firms choosing the latter strategy we have $\left(W_S/B^{1/\omega}\right) = 1$. These two conditions identify unique equilibrium values of W_U, and W_S. Note that at these factor prices firms are indifferent between hiring only skilled workers or only unskilled workers. This indifference guarantees full employment.

APPENDIX B: A NEW DATA SET ON MINCERIAN RETURNS

With Jacopo Ponticelli and Federico Rossi

What is the economic value of an additional year of schooling? How and why does it vary across countries? These questions—the core of the field of labor economics—have received enormous attention in the last few decades. The implications of the answers are obviously far reaching, from the design of educational policy to the evaluation of the importance of human capital as a source of differences in standards of living across countries.

The workhorse empirical model to estimate the returns to education is the human capital earning function introduced by Mincer (1974), where the logarithm of earnings is regressed on years of schooling and a quadratic function of years of experience. This specification has strong theoretical foundations, being the outcome of a standard Ben-Porath (1967) model of human capital accumulation, and, given its simplicity, has been shown to fit the data remarkably well.[1]

In the last few decades, George Psacharopoulos and his co-authors have provided a great service to the profession by compiling extensive collections of estimates of the returns to education for a wide range of countries [Psacharopoulos (1981, 1985, 1994); Psacharopoulos and Patrinos (2004)]. These estimates have been extensively used to analyze cross-country patterns and evaluate the contribution of human capital to economic growth.

The latest available estimates in the aforementioned collections [Psacharopoulos and Patrinos (2004)] are, for most countries,

---

[1] See Card (1999), Heckman et al. (2003), Lemieux (2006), and Polachek (2008) for extensive reviews.

from the 1980s. Since then, however, new studies estimating the returns to education in different countries have burgeoned, thanks to a wealth of new data and econometric techniques which have become available. In this appendix we present a new collection of Mincerian coefficients estimated with data from more recent years. In particular, the data set includes up to two estimates for each country, one for the 1989–1999 period and one for the 2000s; these estimates come from a large number of academic papers and technical reports. We provide a detailed list of sources at http://personal.lse.ac.uk/casellif/.

This appendix is structured as follows. Section B.1 describes the data collection process and the coverage of the data set. Section B.2 offers an overview of the main patterns emerging from the data, and section B.3 concludes.

B.1 Sources and Criteria

In the latest review, Psacharopoulos and Patrinos (2004) emphasize the importance of a selective approach in choosing estimates of returns to education reasonably comparable across countries. In this section we describe the criteria we adopted for the inclusion of an estimate in our data set.

Ideally, we would want to limit ourselves to estimates coming from nationally representative samples, specifications with exactly the same controls, and variables perfectly comparable across countries. Since this would limit our collection to a handful of observations, some compromise is in order to be able to perform meaningful cross-country comparisons.

The estimates included in our data set come from a large number of academic papers and technical reports (see the online appendix for a list of sources). Most of these studies are published in peer-reviewed journals; however, to broaden the coverage we included also unpublished works as long as they met adequate standards in terms of sample size, data quality, and econometric implementation.

To ensure comparability, when selecting the estimates we tried to adhere as closely as possible to the standard Mincerian specification, which includes years of schooling, experience, and experience squared as controls. Many papers we surveyed estimate richer models, controlling for other individual characteristics; luckily for our purposes, results from the baseline specification are often included as well. A particularly common practice is the inclusion of occupational or sectorial dummies: given the occupation is itself an outcome influenced by education, the regression does not have a causal interpretation.[2] We therefore do not include estimates affected by this problem.

Another obstacle for a direct comparison across studies is that exact definition of the dependent variable depends on the context. Whenever possible, we give preference to measures of hourly gross earnings, which are not directly affected by differences in labor supply (part- versus full-time workers) across individuals and in taxation across countries.

As noted by Psacharopoulos and Patrinos (2004), estimates coming from samples of workers employed in the public sector pose additional problems, since their wages are likely not to reflect market wages. We therefore limit ourselves to studies focusing on the private sector.

Finally, as an alternative to the log-linear specification, many papers in the literature estimate models where the returns to schooling are allowed to vary depending on the stage of education. In particular, a common specification consists of regressing the logarithm of earnings on dummies corresponding to the highest level of completed schooling (primary, secondary, and higher) on top of the usual experience controls. As shown in chapter 2, under some assumptions we can establish a one-to-one mapping between these coefficients and the Mincerian return corresponding to the classic log-linear specification. We therefore follow this

---

[2] See Angrist and Pischke (2009) for a detailed discussion of the "bad control" problem.

TABLE B.1. Regional Averages of Mincerian Coefficients

	Year	
Region	1995	2005
Advanced economies	7.79	7.36
Europe and Central Asia	7.37	7.03
Latin America and the Caribbean	10.85	8.17
Africa and Middle East	8.41	8.47
Southeast Asia and the Pacific	8.62	10.58
World	8.70	8.22

method to compute the implied returns and include them whenever an alternative estimate coming from a log-linear specification is not available.

This leaves us with a total of 87 observations for the 1990s and 91 for the 2000s. Many of the countries included in this collection were not previously available, allowing us to provide a more complete picture on the international patterns. Our collection of estimates Mincerian returns is reproduced in Table B.3 at the end of this appendix.

B.2 THE MAIN PATTERNS

The average returns to education by region are shown in table B.1. Overall, the average for all observations included in the data set is 8.70% for the 1990s and 8.22% for the 2000s; these are approximately 1 percentage point lower compared to Psacharopoulos and Patrinos (2004). In terms of regional differences, countries in Latin America and the Caribbean stand out for having the highest returns in 1995, on average just below 11%, while countries in Southeast Asia and the Pacific have the highest returns in 2005; countries in the advanced economies group (as classified by the World Bank) have returns *below* the world average.

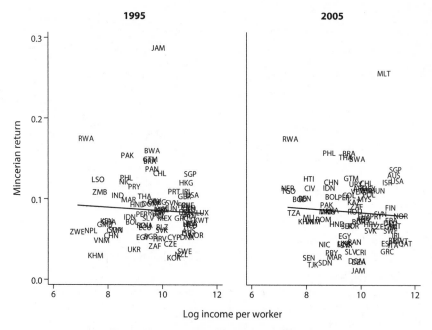

Figure B.1. Mincerian Returns against Income

We now move to consider the correlation between estimated returns and the level of economic development. On a theoretical ground, the relationship is ambiguous: on the one hand, richer countries are endowed with a larger share of educated workers, and if skilled labor is subject to decreasing returns, we should expect lower returns there; on the other hand, the availability of more educated workers could encourage firms to adopt more skill-intensive technologies, widening the productivity gap between skilled and unskilled labor.[3] According to the estimates we collected, there does not appear to be a systematic relationship between returns to schooling and real GDP per capita, either in the 1990s or in the 2000s (figure B.1). Even excluding the two

_____

[3] Moreover, countries' demographic structures and TFP levels might affect the estimated returns, leading to cross-country differences; see Manuelli and Seshadry (2014) for a version of this argument.

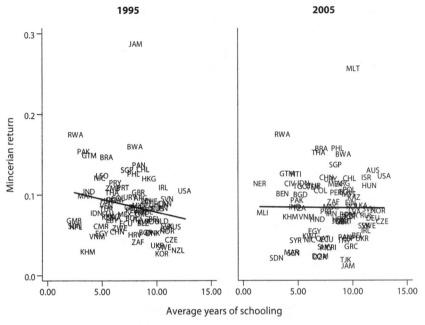

Figure B.2. Mincerian Returns against Years of Schooling

outliers (Jamaica for the 1990s and Malta for the 2000s), the correlation remains slightly negative and not significantly different
from zero at standard confidence levels.

Similar conclusions hold with respect to the relationship with
average years of schooling (figure B.2).[4]

Table B.2 shows the average returns by gender. In both decades
women experience substantially higher returns than men; this is
consistent with the pattern documented in previous collections. In

---

[4] An exception is represented by the downward relationship between returns
to education and average years of schooling in 1995. Excluding the outlier Jamaica,
a regression of Mincerian coefficients on years of schooling (and a constant) yields
an estimated slope of −0.34, significant at the 5% confidence level. Further, using
an extended version of the data set constructed by Psacharopoulos and Patrinos
(2004), Banerjee and Duflo (2005) find a small but significant negative relationship between Mincerian returns and both GDP per capita and average years of
schooling.

recent work, Pitt et al. (2012) document that this gap in returns to schooling cannot simply be ascribed to differences in the quantity of education across genders, since in most countries women have higher educational attainment than men. They instead propose an explanation based on comparative advantage due to biological differences in the endowment of skill and brawn.

TABLE B.2. Average Mincerian
Coefficients across Genders

Gender	Year	
	1995	2005
Men	9.12	7.82
Women	10.01	9.52

TABLE B.3. Mincerian Coefficients

Country	1990s	2000s
Algeria		2.2
Argentina	9.6	11.4
Australia	6	13
Austria	7.9	6.7
Bangladesh		10
Belgium	7	4.9
Belize	6.5*	
Benin		10.1*
Bolivia	7.1	10.3
Bosnia and Herzegovina		9
Botswana	16	15
Brazil	14.7	15.7
Bulgaria	5.3	7.2
Cambodia	2.9	7.2
Cameroon	6.1*	
Canada	8.9	
Chile	13.2	12

TABLE B.3. (Continued)

Country	1990s	2000s
China	5.4	12.1
Colombia	9.6	10.5
Costa Rica	8.5	3.4*
Côte d'Ivoire		11.4*
Croatia	5	6.9
Cyprus	5.2	
Czech Republic	4.4	6.6
Denmark	5.3	
Dominican Republic	9.4	2.3*
Ecuador	6.4*	4.4*
Egypt	5.2	5.4
El Salvador	7.6	3.5*
Eritrea		11
Ethiopia	14.7	
Finland	9.2	9
France	7.8	7.4
Gambia	6.8	
Georgia		1.1
Germany	8.7	7
Ghana	7.1	
Greece	7.6	3.5
Guatemala	14.9	12.6
Haiti		12.6*
Honduras	9.3	6.9*
Hong Kong	12	
Hungary	8.8	11.1
India	10.4	8.5
Indonesia	7.8	11.4
Iran		7.6
Ireland	10.9	5.5
Israel	5.7	12.1
Italy	6.9	4.3
Jamaica	28.8	1.1*

(continued)

TABLE B.3. Mincerian Coefficients (Continued)

Country	1990s	2000s
Japan	8.3	
Jordan		6.7
Kazakhstan		9.6
Kenya	7.3*	
Kuwait	7.3	4.8
Latvia	6.7	7.8
Lesotho	12.4	
Libya	6.8	
Luxembourg	8.3	
Macedonia		5.9
Malaysia	8.7*	10
Mali		7.7*
Malta		25.7
Mexico	7.6	11.3
Moldova		7.5
Mongolia		8.5
Morocco	9.9	2.8
Nepal	6	
Netherlands	6.7	
New Zealand	3.1	
Nicaragua	12.1	4.4*
Niger		11.4*
Nigeria	4.8	4.5
Northern Ireland	16	
Norway	5.5	7.9
Pakistan	15.4	9.3
Palestine		5.4
Panama	13.7	4.7*
Paraguay	11.5	3.3*
Peru	8.1	10.3
Philippines	12.6	15.8
Poland	7.9	10.6
Portugal	10.9	7.9

TABLE B.3. (Continued)

Country	1990s	2000s
Qatar		4.5
Romania	6.7	8.5
Russia	8.3	7.4
Rwanda	17.5	17.5
Senegal		2.7
Serbia		6.7
Singapore	13.1	13.7
Slovak Republic	6.1	6.1
Slovenia	9.5	8.2
South Africa	4.1	9.1
South Korea	2.7	
Spain	8.4	4.5
Sri Lanka		8.6
Sudan	6.1	2.1
Sweden	3.5	6.1
Switzerland	9.2	
Syria		4.3*
Tajikistan		1.8
Tanzania		8.3
Thailand	10.3	15.2
Togo		11*
Turkey	8.3	11.1
Ukraine	3.7	4.5
United Kingdom	10.3	6.6
United States	10.5	12.3
Uruguay	9.7	11.9
Venezuela	8.7	11
Vietnam	4.8	7.2
Zambia	10.9	
Zimbabwe	5.9*	

Notes: Superscript * indicates that the Mincerian return was constructed from the secondary school premium, as described in the text.

REFERENCES

Acemoglu, D. (1998): "Why Do New Technologies Complement Skills? Directed Technical Change and Wage Inequality," *Quarterly Journal of Economics*, vol. 113(4), pp. 1055–89.

Acemoglu, D. (2002): "Directed Technical Change," *Review of Economic Studies*, vol. 69(4), pp. 781–809.

Acemoglu, D., and D. H. Autor (2011): "Skills, Tasks and Technologies: Implications for Employment and Earnings," in Ashenfelter, O. and D. Card, eds. *Handbook of Labor Economics,* vol. 4, Elsevier.

Acemoglu, D., and F. Zilibotti (2001): "Productivity Differences," *Quarterly Journal of Economics*, vol. 116(2), pp. 563–606.

Angrist, J. D., and J. S. Pischke (2009): *Mostly Harmless Econometrics: An Empiricist's Companion*, Princeton University Press.

Antràs, P. (2004): "Is the U.S. Aggregate Production Function Cobb-Douglas? New Estimates of the Elasticity of Substitution," *Contributions in Macroeconomics*, vol. 4 (1).

Atkinson, A. B., and J. E. Stiglitz (1969): "A New View of Technological Change," *Economic Journal*, vol. 79(315), pp. 573–78.

Autor, D. H., L. F. Katz, and A. B. Krueger (1998): "Computing Inequality: Have Computers Changed the Labor Market?" *Quarterly Journal of Economics*, vol. 113(4), pp. 1169–1213.

Banerjee, A. V., and E. Duflo (2005): "Growth Theory through the Lens of Development Economics," in Aghion, P., and S. Durlauf, eds., *Handbook of Economic Growth*, vol. 1, chapter 7, pp. 473–552, Elsevier.

Barro, R. J., and W. L. Lee (2013): "A New Data Set of Educational Attainment in the World, 1950–2010," *Journal of Development Economics,* vol. 104, pp. 184–98.

Basu, S., and D. N. Weil (1998): "Appropriate Technology and Growth," *Quarterly Journal of Economics*, vol. 113(4), pp. 1025–54.

Ben-Porath, Y. (1967): "The Production of Human Capital and the Life Cycle of Earnings," *Journal of Political Economy*, vol. 75(4), pp. 352–65.

Bernanke, B., and R. S. Gurkaynak (2001): "Is Growth Exogenous? Taking Mankiw, Romer, and Weil Seriously," in Bernanke, B., and K. S. Rogoff, eds., *NBER Macroeconomics Annual 2001*, vol.16, pp. 11–57, MIT Press.

Bils, M., and P. Klenow (2000): "Does Schooling Cause Growth?" *American Economic Review*, vol. 90, pp. 1160–83.

Boehm, M., and C. Siegel (2014): "The Race between the Demand and Supply of Experience," PDF slides.

Card, D. (1999): "The Causal Effect of Education on Earnings," in Ashenfelter, O., and D. Card, eds., *Handbook of Labor Economics*, vol. 3, pp. 1801–63, Elsevier.

Card, D., and T. Lemieux (2001): "Can Falling Supply Explain the Rising Return to College for Younger Men? A Cohort-Based Analysis," *Quarterly Journal of Economics*, vol. 116(2), pp. 705–46.

Caselli, F. (1999): "Technological Revolutions," *American Economic Review*, vol. 89(1), pp. 78–102.

Caselli, F. (2005): "Accounting for Cross-Country Income Differences," in Aghion, P., and S. Durlauf, eds., *Handbook of Economic Growth*, vol. 1, pp. 679–741, Elsevier.

Caselli, F. (2008a): "Growth Accounting," *Palgrave Dictionary of Economics*.

Caselli, F. (2008b): "Level Accounting," *Palgrave Dictionary of Economics*.

Caselli, F., and A. Ciccone (2013): "The Contribution of Schooling in Development Accounting: Results from a Nonparametric Upper Bound," *Journal of Development Economics*, vol. 104(C), pp. 199–211.

Caselli, F., and W. J. Coleman (2001): "Cross-Country Technology Diffusion: The Case of Computers," *American Economic Review*, vol. 91(2), pp. 328–35.

Caselli, F., and W. J. Coleman (2002): "The U.S. Technology Frontier," *American Economic Review*, vol. 92(2), pp. 148–52.

Caselli, F., and W. J. Coleman (2006): "The World Technology Frontier," *American Economic Review*, vol. 96(3), pp. 499–522.

Caselli, F., and J. Feyrer (2007): "The Marginal Product of Capital," *Quarterly Journal of Economics*, vol. 122(2), pp. 535–68.

Caselli, F., and D. J. Wilson (2004): "Importing Technology," *Journal of Monetary Economics*, vol. 51(1), pp. 1–32.

Ciccone, A., and G. Peri (2005): "Long-Run Substitutability between More and Less Educated Workers: Evidence from U.S. States 1950–1990," *Review of Economics and Statistics*, vol. 87(4), pp. 652–63.

Cohen, D., and M. Soto (2007): "Growth and Human Capital: Good Data, Good Results," *Journal of Economic Growth*, vol. 12(1), pp. 51–76.

De Philippis, M., and F. Rossi (2016): "Parents, Schools and Human Capital Differences across Countries," unpublished, London School of Economics.

Diamond, P., D. L. McFadden, and M. Rodriquez (1978): "Measurement of the Elasticity of Factor Substitution and Bias of Technical Change," in Fuss, M., and D. L. McFadden, eds., *Production Economics: A Dual*

Approach to Theory and Applications. Volume II: Applications to the Theory of Production, chapter 5, North-Holland.

Diwan, I., and D. Rodrik (1991): "Patents, Appropriate Technology, and North-South Trade," *Journal of International Economics*, vol. 30(1–2), pp. 27–47.

Duffy, J., and C. Papageorgiou (2000): "A Cross-Country Empirical Investigation of the Aggregate Production Function Specification," *Journal of Economic Growth*, vol. 5(1), pp. 87–120.

Elsby, M. W., B. Hobijn, and A. Sahin (2013): "The Decline of the U.S. Labor Share," *Brookings Papers on Economic Activity*, vol. (2), pp. 1–63.

Goldin, C., and L. F. Katz (2008): *The Race between Education and Technology*, Harvard University Press.

Gollin, D. (2002): "Getting Income Shares Right," *Journal of Political Economy*, vol. 110(2), pp. 458–74.

Gundlach, E., L. Woessmann, and J. Gmelin (2001): "The Decline of Schooling Productivity in OECD Countries," *Economic Journal*, vol. 111, pp. C135–C147.

Guvenen, F., and B. Kuruscu (2010): "A Quantitative Analysis of the Evolution of the U.S. Wage Distribution: 1970–2000," *NBER Macroeconomics Annual 2009*, vol. 24, pp. 227–276, MIT Press.

Hall, Robert E., and Charles I. Jones (1996): "The Productivity of Nations," NBER Working Paper No. 5812.

Hamermesh, D. S. (1986): "The Demand for Labor in the Long Run," in Ashenfelter, O., and R. Layard, eds., *Handbook of Labor Economics*, vol. 1, Elsevier.

Hanushek, E., and D. D. Kimko (2000): "Schooling, Labor-Force Quality, and the Growth of Nations," *American Economic Review*, vol. 90(5), pp. 1184–1208.

Hanushek, E., and L. Woessmann (2012): "Do Better Schools Lead to More Growth? Cognitive Skills, Economic Outcomes, and Causation," *Journal of Economic Growth*, vol. 17(4), pp. 267–321.

Hanushek, E., and L. Zhang (2009): "Quality-Consistent Estimates of International Schooling and Skill Gradients," *Journal of Human Capital*, vol. 3(2), pp. 107–43.

Heckman, J. J., L. J. Lochner, and P. E. Todd (2003): "Fifty Years of Mincer Earnings Regressions," NBER Working Paper No. 9732.

Heston, A., R. Summers, and B. Aten (2012): Penn World Table Version 7.1, Centre for International Comparisons of Production, Income and Prices at the University of Pennsylvania.

Hicks, J. R. (1932): *The Theory of Wages*, MacMillan.

Hicks, J. R. (1939): *Value and Capital*, 2nd ed., Oxford University Press.

Hsieh, C., and P. J. Klenow (2007): "Relative Prices and Relative Prosperity," *American Economic Review*, vol. 97(3), pp. 562–85.

Jeong, H., Y. Kim, and I. Manovskii (2015): "The Price of Experience," *American Economic Review*, vol. 105(2).

Jones, C. I. (2005): "The Shape of Production Functions and the Direction of Technical Change," *Quarterly Journal of Economics*, vol. 120(2), pp. 517–49.

Katz, L. F., and D. H. Autor (1999): "Changes in the Wage Structure and Earnings Inequality," in Ashenfelter, O., and D. Card, eds., *Handbook of Labor Economics*, vol. 3A, pp. 1463–555, Elsevier.

Katz, L. F., and K. M. Murphy (1992): "Changes in Relative Wages, 1963–1987: Supply and Demand Factors," *Quarterly Journal of Economics*, vol. 107(1), pp. 35–78.

Kennedy, C. (1964): "Induced Bias in Innovation and the Theory of Distribution," *Economic Journal*, vol. 74, pp. 541–47.

Krusell, P., L. Ohanian, V. Rios-Rull, and G. Violante (2000): "Capital–Skill Complementarity and Inequality: A Macroeconomic Analysis," *Econometrica*, vol. 68, pp. 1029–53.

Lagakos, D., B. Moll, T. Porzio, N. Qian, and T. Schoellman (2012): "Experience Matters: Human Capital and Development Accounting," NBER Working Paper No. 18602.

Lemieux, T. (2006): "The "Mincer Equation: Thirty Years after *Schooling, Experience, and Earnings*," in Grossbard, S., ed., *Jacob Mincer: A Pioneer of Modern Labor Economics*, pp. 127–45, Springer.

Manuelli, R., and A. Seshadri (2014): "Human Capital and the Wealth of Nations," *American Economic Review*, vol. 104(9), pp. 2736–62.

Mincer, J. A. (1974): *Schooling, Experience, and Earnings*, Columbia University Press.

Monge-Naranjo, A., J. M. Sanchez, and R. Santaeulalia-Llopis (2015): "Natural Resources and Global Misallocation," Federal Reserve Bank of St. Louis, unpublished.

Neiman, B., and L. Karabarbounis (2014): "The Global Decline of the Labor Share," *Quarterly Journal of Economics* , vol. 129(1), pp. 61–103.

Oberfield, E., and D. Raval (2012): "Micro Data and the Macro Elasticity of Substitution," U.S. Census Bureau Center for Economic Studies Paper No. CES-WP-12-05.

O'Mahony, M., and P. Timmer (2009): "Output, Input and Productivity Measures at the Industry Level: The EU KLEMS Database," *Economic Journal*, vol. 119, pp. F374–403.

Pitt, M., M. Rosenzweig, and M. N. Hassan (2012): "Human Capital Investment and the Gender Division of Labor in a Brawn-Based Economy," *American Economic Review*, vol. 102(7), pp. 3531–60.

Polachek, S. W. (2008): "Earnings over the Life Cycle: The Mincer Earnings Function and Its Applications," *Foundations and Trends(R) in Microeconomics*, vol. 4 (3), pp. 165–272.

Psacharopoulos, G. (1981): "Returns to Education: An Updated International Comparison," *Comparative Education*, vol. 17(3), pp. 321–41.

Psacharopoulos, G. (1985): "Returns to Education: A Further International Update and Implications," *Journal of Human Resources*, vol. 20(4), pp. 583–604.

Psacharopoulos, G. (1994): "Returns to Investment in Education: A Global Update," *World Development*, vol. 22(9), pp. 1325–43.

Psacharopoulos, G., and H. A. Patrinos (2004): "Returns to Investment in Education: A Further Update," *Education Economics*, vol. 12(2), pp. 111–34.

Ruiz-Arranz, M. (2002): "Wage Inequality in the U.S.: Capital–Skill Complementarity vs. Skill-Biased Technological Change," unpublished.

Samuelson, P. A. (1965): "A Theory of Induced Innovation along Kennedy–Weisacker Lines," *Review of Economics and Statistics*, vol. 47(4), pp. 343–56.

Samuelson, P. A. (1966): "Rejoinder: Agreements, Disagreements, Doubts, and the Case of Induced Harrod-Neutral Technical Change," *Review of Economics and Statistics*, vol. 48(4), pp. 444–48.

Ventura, J. (1997): "Growth and Interdependence," *Quarterly Journal of Economics*, vol. 112(1), pp. 57–84.

Weil, D. (2007): "Accounting for the Effect of Health on Economic Growth," *Quarterly Journal of Economics*, vol. 122, pp. 1265–1306.

World Bank (2011): *The Changing Wealth of Nations: Measuring Sustainable Development in the New Millennium*.

World Bank (2012): *World Development Indicators*.

INDEX

Italicized pages refer to figures and tables.